Basic Computation

Working with Percents

Loretta M. Taylor, Ed. D.
Mathematics Teacher
Hillsdale High School
San Mateo, California

Harold D. Taylor, Ed. D.
Head, Mathematics Department
Aragon High School
San Mateo, California

DALE
SEYMOUR
PUBLICATIONS
P.O. BOX 10888
PALO ALTO, CA 94303

Editor: Elaine C. Murphy
Production Coordinator: Ruth Cottrell
Cover Designer: Michael Rogondino
Compositor: WB Associates
Printer: Malloy Lithographing

ISBN 0-86651-004-4

Catalog order number DS01185

fghi-MA-9043210

DALE
SEYMOUR
PUBLICATIONS
P.O. BOX 10888
PALO ALTO, CA 94303

ABOUT THE PROGRAM

WHAT IS THE BASIC COMPUTATION LIBRARY?

The books in the BASIC COMPUTATION library together provide comprehensive practice in all the essential computational skills. There are practice books and a test book. The practice books consist of carefully sequenced drill worksheets organized in groups of five. The test book contains daily quizzes (160 quizzes in all), semester tests, and year-end tests written in standardized-test formats.

If you find this book effective, you may want to use others in the series. Build your own library to suit your own needs.

BOOK 1	WORKING WITH WHOLE NUMBERS
BOOK 2	UNDERSTANDING FRACTIONS
BOOK 3	WORKING WITH FRACTIONS
BOOK 4	WORKING WITH DECIMALS
BOOK 5	WORKING WITH PERCENTS
BOOK 6	UNDERSTANDING MEASUREMENT
BOOK 7	FINDING AREA AND PERIMETER
BOOK 8	WORKING WITH CIRCLES AND VOLUME
BOOK 9	APPLYING COMPUTATIONAL SKILLS
TEST BOOK	BASIC COMPUTATION QUIZZES AND TESTS

WHO CAN USE THE BASIC COMPUTATION LIBRARY?

Classroom teachers, substitute teachers, tutors, parents, and persons wishing to study on their own can use these materials. Although written specifically for the general math classroom, books in the BASIC COMPUTATION library can be used with any program requiring carefully sequenced computational practice. The material is appropriate for use with any person, young or old, who has not yet certified computational proficiency. It is especially suitable for middle school, junior high school, and high school students who need to master the essential computational skills necessary for mathematical literacy.

WHAT IS IN THIS BOOK?

This book is a practice book. In addition to these teacher notes, it contains student worksheets, example problems, and a record form.

Worksheets

The worksheets are designed to give even the slowest student a chance to master the essential computational skills. Most worksheets come in five equivalent forms allowing for pretesting, practice, and posttesting on any one skill. Each set of worksheets provides practice in only one or two specific skills and the work progresses in very small steps from one set to the next. Instructions are clear and simple, with handwritten samples of the exercises completed. Ample practice is provided on each page, giving students the opportunity to strengthen their skills. Answers to each problem are included in the back of the book.

Example Problems

Fully-worked examples show how to work each type of exercise. Examples are keyed to the worksheet pages. The example solutions are written in a straightforward manner and are easily understood.

Record Form

A record form is provided to help in recording progress and assessing instructional needs.

Answers

Answers to each problem are included in the back of the book.

HOW CAN THE BASIC COMPUTATION LIBRARY BE USED?

The materials in the BASIC COMPUTATION library can serve as the major skeleton of a skills program or as supplements to any other computational skills program. The large number of worksheets gives a wide variety from which to choose and allows flexibility in structuring a program to meet individual needs. The following suggestions are offered to show how the BASIC COMPUTATION library may be adapted to a particular situation.

Minimal Competency Practice

In various fields and schools, standardized tests are used for entrance, passage from one level to another, and certification of competency or proficiency prior to graduation. The materials in the BASIC COMPUTATION library are particularly well-suited to preparing for any of the various mathematics competency tests, including the mathematics portion of the General Educational Development test (GED) used to certify high school equivalency.

Together, the books in the BASIC COMPUTATION library give practice in all the essential computational skills measured on competency tests. The semester tests and year-end tests from the test book are written in standardized-test formats. These tests can be used as sample minimal competency tests. The worksheets can be used to brush up on skills measured by the competency tests.

Skill Maintenance

Since most worksheets come in five equivalent forms, the computation work can be organized into weekly units as suggested by the following schedule. Day one is for pretesting and introducing a skill. The next three days are for drill and practice followed by a unit test on the fifth day.

AUTHOR'S SUGGESTED TEACHING SCHEDULE

	Day 1	Day 2	Day 3	Day 4	Day 5
Week 1	pages 1 and 2 pages 11 and 12	pages 3 and 4 pages 13 and 14	pages 5 and 6 pages 15 and 16	pages 7 and 8 pages 17 and 18	pages 9 and 10 pages 19 and 20
Week 2	pages 21 and 22 pages 31 and 32	pages 23 and 24 pages 33 and 34	pages 25 and 26 pages 35 and 36	pages 27 and 28 pages 37 and 38	pages 29 and 30 pages 39 and 40
Week 3	pages 41 and 42	pages 43 and 44	pages 45 and 46	pages 47 and 48	pages 49 and 50
Week 4	pages 51 and 52 page 61	pages 53 and 54 page 62	pages 55 and 56 page 63	pages 57 and 58 page 64	pages 59 and 60 page 65

The daily quizzes from BASIC COMPUTATION QUIZZES AND TESTS can be used on the drill and practice days for maintenance of previously-learned skills or diagnosis of skill deficiencies.

A five-day schedule can begin on any day of the week. The author's ideal schedule begins on Thursday, with reteaching on Friday. Monday and Tuesday are for touch-up teaching and individualized instruction. Wednesday is test day.

Supplementary Drill

There are more than 18,000 problems in the BASIC COMPUTATION library. When students need more practice with a given skill, use the appropriate worksheets from the library. They are suitable for classwork or homework practice following the teaching of a specific skill. With five equivalent pages for most worksheets, adequate practice is provided for each essential skill.

HOW ARE MATERIALS PREPARED?

The books are designed so the pages can be easily removed and reproduced by Thermofax, Xerox, or a similar process. For example, a ditto master can be made on a Thermofax for use on a spirit duplicator. Permanent transparencies can be made by processing special transparencies through a Thermofax or Xerox.

Any system will run more smoothly if work is stored in folders. Record forms can be attached to the folders so that either students or teachers can keep records of individual progress. Materials stored in this way are readily available for conferences.

EXAMPLE PROBLEMS

PERCENT TO FRACTION AND DECIMAL

EXAMPLE 1 Rewrite 57% as a fraction with a denominator of 100.

Solution: $57\% = \frac{57}{100}$

EXAMPLE 2 Rewrite 45% as a fraction in lowest terms.

Solution:

Method 1: $45\% = \frac{45}{100} = \frac{3 \cdot 3 \cdot 5}{2 \cdot 2 \cdot 5 \cdot 5} = \frac{3 \cdot 3}{2 \cdot 2 \cdot 5} = \frac{9}{20}$

Method 2: $45\% = \frac{45}{100} = \frac{45 \div 5}{100 \div 5} = \frac{9}{20}$

EXAMPLE 3 Rewrite 225% and $37\frac{1}{2}\%$ in decimal form.

Solution:

Method 1: Omit the percent sign and move the decimal point two places to the *left*.

$225\% = 225.\% = 2.25$

$37\frac{1}{2}\% = 37.5\% = 0.375$

Method 2: $225\% = \frac{225}{100} = 2.25$

$37\frac{1}{2}\% = \frac{37.5}{100} = 0.375$

FRACTION TO DECIMAL AND PERCENT

EXAMPLE 1 Rewrite $\frac{84}{100}$ as a fraction in lowest terms.

Solution:

Method 1: $\frac{84}{100} = \frac{2 \cdot 2 \cdot 3 \cdot 7}{2 \cdot 2 \cdot 5 \cdot 5} = \frac{3 \cdot 7}{5 \cdot 5} = \frac{21}{25}$

Method 2: $\frac{84}{100} = \frac{84 \div 4}{100 \div 4} = \frac{21}{25}$

EXAMPLE 2 Rewrite $\frac{35}{100}$ in decimal form.

Solution: $\frac{35}{100} = 0.35$

EXAMPLE 3 Rewrite $\frac{13}{100}$ and $\frac{185}{100}$ in percent form.

Solution: When the denominator of a fraction is 100, replace the fraction bar and 100 with a percent sign.

$\frac{13}{100} = 13\%$

$\frac{185}{100} = 185\%$

FRACTION OF A NUMBER; PERCENT OF A NUMBER

EXAMPLE 1 Find $\frac{4}{5}$ of 110.

Solution: $\frac{4}{5}$ of 110 means $\frac{4}{5} \times 110$.

$$\frac{4}{5} \times 110 = \frac{4}{\overset{}{\underset{1}{5}}} \times \frac{\overset{22}{110}}{1} = \frac{88}{1} = 88$$

EXAMPLE 2 Find 80% of 110.

Solution: 80% of 110 means 80% × 110.

$$80\% \times 110 = 0.80 \times 110 = 88$$

FRACTION OF A NUMBER; FINDING PERCENTAGE

EXAMPLE 1 Find $\frac{7}{5}$ of 85.

Solution: $\frac{7}{5}$ of 85 means $\frac{7}{5} \times 85$.

$$\frac{7}{5} \times 85 = \frac{7}{\overset{}{\underset{1}{5}}} \times \frac{\overset{17}{85}}{1} = \frac{119}{1} = 119$$

EXAMPLE 2 Find 163% of 74.

Solution: 163% of 74 means 163% × 74.

$$163\% \times 74 = 1.63 \times 74 = 120.62$$

FRACTION TO PERCENT; FINDING PERCENTAGE

EXAMPLE 1 Rewrite $\frac{3}{8}$ as a percent.

Solution:

Method 1: $\frac{3}{8} = 3 \div 8 = 0.375 = 37.5\%$

Method 2: $\frac{3}{8} = \frac{3}{8} \cdot \frac{12\frac{1}{2}}{12\frac{1}{2}} = \frac{37\frac{1}{2}}{100} = 37\frac{1}{2}\%$

EXAMPLE 2 Rewrite $\frac{1}{3}$ as a percent.

Solution:

Method 1: When division results in a repeating decimal, carry the division to the hundredths place and use the remainder to make a fraction.

$$\frac{1}{3} = 1 \div 3 = 0.33\frac{1}{3} = 33\frac{1}{3}\%$$

Method 2: $\frac{1}{3} = \frac{1}{3} \cdot \frac{33\frac{1}{3}}{33\frac{1}{3}} = \frac{33\frac{1}{3}}{100} = 33\frac{1}{3}\%$

EXAMPLE 3 Find 75% of 32.

Solution: 75% of 32 means 75% × 32.

$$75\% \times 32 = 0.75 \times 32 = 24$$

EXAMPLE 4 Find $33\frac{1}{3}$% of 42.

Solution: $33\frac{1}{3}$% of 42 means $33\frac{1}{3}$% × 42.

$33\frac{1}{3}$% × 42 = $\frac{1}{3}$ × 42 = 14

PERCENT TO DECIMAL; FINDING PERCENTAGE

EXAMPLE 1 Rewrite 0.7% in decimal form.

Solution: Omit the percent sign and move the decimal point two places to the left.

0.7% = 000.7% = 0.007

EXAMPLE 2 Find 150% of 21.

Solution: 150% of 21 means 150% × 21.

150% × 21 = 1.50 × 21 = 31.5

PERCENT OF A NUMBER

EXAMPLE 1 Find 12.6% of 32.

Solution: 12.6% of 32 means 12.6% × 32.

12.6% × 32 = 0.126 × 32 = 4.032

FINDING PERCENT

EXAMPLE 1 1 is what percent of 4?

Solution: rate = percentage ÷ base

= 1 ÷ 4

= 25%

SALES

EXAMPLE 1 Find the savings and sale price of an item regularly sold for $24.95 and marked for a 20% discount.

Solution: savings = regular price × discount

= $24.95 × 20%

= $24.95 × 0.20

= $4.99

sale price = regular price − savings

= $24.95 − $4.99

= $19.96

SAVINGS AND SALE PRICE

EXAMPLE 1 Find the savings and sale price of a pair of jeans regularly sold for $25.00 and marked for a 15% discount.

Solution: savings = regular price × discount
= $25.00 × 15%
= $25.00 × 0.15
= $3.75

sale price = regular price − savings
= $25.00 − $3.75
= $21.25

FRACTION TO PERCENT

EXAMPLE 1 Rewrite $\frac{4}{5}$ as a percent.

Solution: $\frac{4}{5}$ = 4 ÷ 5 = 0.20 = 20%

EXAMPLE 2 4 is what percent of 5?

Solution: rate = percentage ÷ base
= 4 ÷ 5
= 20%

FINDING THE WHOLE; FINDING PERCENTAGE

EXAMPLE 1 27 is 75% of what number?

Solution: base = percentage ÷ rate
= 27 ÷ 75%
= 27 ÷ 0.75
= 36

EXAMPLE 2 43.8% of 126 is what number?

Solution: 43.8% of 126 means 43.8% × 126.
43.8% × 126 = 0.438 × 126
= 55.188

UNDERSTANDING SALES

EXAMPLE 1 Use the advertisement to find the regular price, sale price, amount saved, and percent discount.

SAVE ON RECORDERS!

regular $24.00   now $19.20

Solution: regular price = $24.00

sale price = $19.20

amount saved = regular price − sale price

= $24.00 − $19.20

= $4.80

percent discount = amount saved ÷ regular price

= $4.80 ÷ $24.00

= 0.20

= 20%

STUDENT RECORD SHEET

Worksheets Completed

Page Number

1	3	5	7		9
2	4	6	8		10
11	13	15	17		19
12	14	16	18		20
21	23	25	27		29
22	24	26	28		30
31	33	35	37		39
32	34	36	38		40
41	43	45	47		49
42	44	46	48		50
51	53	55	57		59
52	54	56	58		60
61	62	63	64		65

Daily Quiz Grades

No.	Score

Check List Skill Mastered

		Date
☐	percent to fraction	
☐	percent to decimal	
☐	fraction to decimal	
☐	fraction to percent	
☐	decimal to percent	
☐	fraction of a number	
☐	percent of a number (using fraction)	
☐	percent of a number (using decimal)	
☐	finding percent	
☐	finding savings from sale	
☐	finding sales price	
☐	finding the whole	
☐	reading sales information from ads	
☐	finding amount saved from sale	
☐	finding percent saved from sale	

Notes

_____

_____

_____

_____

_____

_____

_____

_____

Percent to fraction and decimal   Name _____

                                       Date _____

Rewrite the percents in three ways.

	percent form	fraction (denominator of 100)	fraction (lowest terms)	decimal form
1.	24%	$\frac{24}{100}$	$\frac{6}{25}$	0.24
2.	56%			
3.	92%			
4.	105%			
5.	69%			
6.	53%			
7.	28%			
8.	75%			
9.	25%			
10.	50%			
11.	$33\frac{1}{3}$%			
12.	32%			
13.	145%			
14.	$62\frac{1}{2}$%			
15.	176%			
16.	225%			
17.	95%			
18.	42%			
19.	265%			
20.	67%			

Name _____

Date _____

Rewrite the fractions in three ways.

	fraction (denominator of 100)	fraction (lowest terms)	decimal form	percent form
1.	$\frac{50}{100}$	$\frac{1}{2}$	0.50	50%
2.	$\frac{25}{100}$			
3.	$\frac{175}{100}$			
4.	$\frac{140}{100}$			
5.	$\frac{240}{100}$			
6.	$\frac{170}{100}$			
7.	$\frac{55}{100}$			
8.	$\frac{135}{100}$			
9.	$\frac{185}{100}$			
10.	$\frac{204}{100}$			
11.	$\frac{125}{100}$			
12.	$\frac{12}{100}$			
13.	$\frac{56}{100}$			
14.	$\frac{88}{100}$			
15.	$\frac{96}{100}$			
16.	$\frac{144}{100}$			
17.	$\frac{176}{100}$			
18.	$\frac{212}{100}$			
19.	$\frac{20}{100}$			
20.	$\frac{45}{100}$			

Percent to fraction and decimal

Rewrite the percents in three ways.

	percent form	fraction (denominator of 100)	fraction (lowest terms)	decimal form
1.	68%	$\frac{68}{100}$	$\frac{17}{25}$	0.68
2.	73%			
3.	46%			
4.	29%			
5.	50%			
6.	34%			
7.	26%			
8.	81%			
9.	49%			
10.	56%			
11.	59%			
12.	98%			
13.	156%			
14.	216%			
15.	44%			
16.	114%			
17.	29%			
18.	32%			
19.	67%			
20.	$37\frac{1}{2}\%$			

Fraction to decimal and percent

Name _____

Date _____

Rewrite the fractions in three ways.

	fraction (denominator of 100)	fraction (lowest terms)	decimal form	percent form
1.	$\frac{245}{100}$	$\frac{49}{20}$	2.45	245%
2.	$\frac{85}{100}$			
3.	$\frac{180}{100}$			
4.	$\frac{16}{100}$			
5.	$\frac{120}{100}$			
6.	$\frac{270}{100}$			
7.	$\frac{304}{100}$			
8.	$\frac{155}{100}$			
9.	$\frac{35}{100}$			
10.	$\frac{70}{100}$			
11.	$\frac{225}{100}$			
12.	$\frac{150}{100}$			
13.	$\frac{132}{100}$			
14.	$\frac{176}{100}$			
15.	$\frac{32}{100}$			
16.	$\frac{52}{100}$			
17.	$\frac{128}{100}$			
18.	$\frac{20}{100}$			
19.	$\frac{30}{100}$			
20.	$\frac{230}{100}$			

Percent to fraction and decimal   Name _____

                                          Date _____

Rewrite the percents in three ways.

	percent form	fraction (denominator of 100)	fraction (lowest terms)	decimal form
1.	35%	$\frac{35}{100}$	$\frac{7}{20}$	0.35
2.	60%			
3.	42%			
4.	28%			
5.	39%			
6.	$37\frac{1}{2}\%$			
7.	40%			
8.	76%			
9.	53%			
10.	75%			
11.	65%			
12.	$33\frac{1}{3}\%$			
13.	72%			
14.	150%			
15.	47%			
16.	12%			
17.	175%			
18.	67%			
19.	$12\frac{1}{2}\%$			
20.	85%			

Name _____

Date _____

Rewrite the fractions in three ways.

	fraction (denominator of 100)	fraction (lowest terms)	decimal form	percent form
1.	$\frac{80}{100}$	$\frac{4}{5}$	0.80	80%
2.	$\frac{95}{100}$			
3.	$\frac{255}{100}$			
4.	$\frac{110}{100}$			
5.	$\frac{105}{100}$			
6.	$\frac{215}{100}$			
7.	$\frac{168}{100}$			
8.	$\frac{112}{100}$			
9.	$\frac{44}{100}$			
10.	$\frac{136}{100}$			
11.	$\frac{275}{100}$			
12.	$\frac{25}{100}$			
13.	$\frac{5}{100}$			
14.	$\frac{84}{100}$			
15.	$\frac{188}{100}$			
16.	$\frac{225}{100}$			
17.	$\frac{250}{100}$			
18.	$\frac{160}{100}$			
19.	$\frac{65}{100}$			
20.	$\frac{220}{100}$			

Percent to fraction and decimal   Name _____

                                         Date _____

Rewrite the percents in three ways.

	percent form	fraction (denominator of 100)	fraction (lowest terms)	decimal form
1.	102%	$\frac{102}{100}$	$\frac{51}{50}$	1.02
2.	53%			
3.	30%			
4.	36%			
5.	69%			
6.	20%			
7.	5%			
8.	38%			
9.	25%			
10.	84%			
11.	62%			
12.	115%			
13.	48%			
14.	34%			
15.	81%			
16.	$66\frac{2}{3}\%$			
17.	325%			
18.	$37\frac{1}{2}\%$			
19.	155%			
20.	$12\frac{1}{2}\%$			

Fraction to decimal and percent

Name _____

Date _____

Rewrite the fractions in three ways.

	fraction (denominator of 100)	fraction (lowest terms)	decimal form	percent form
1.	$\frac{20}{100}$	$\frac{1}{5}$	0.20	20%
2.	$\frac{140}{100}$			
3.	$\frac{150}{100}$			
4.	$\frac{210}{100}$			
5.	$\frac{145}{100}$			
6.	$\frac{115}{100}$			
7.	$\frac{76}{100}$			
8.	$\frac{176}{100}$			
9.	$\frac{8}{100}$			
10.	$\frac{75}{100}$			
11.	$\frac{175}{100}$			
12.	$\frac{90}{100}$			
13.	$\frac{180}{100}$			
14.	$\frac{205}{100}$			
15.	$\frac{148}{100}$			
16.	$\frac{196}{100}$			
17.	$\frac{22}{100}$			
18.	$\frac{24}{100}$			
19.	$\frac{240}{100}$			
20.	$\frac{165}{100}$			

Percent to fraction and decimal

Name _____

Date _____

Rewrite the percents in three ways.

	percent form	fraction (denominator of 100)	fraction (lowest terms)	decimal form
1.	28%	$\frac{28}{100}$	$\frac{7}{25}$	0.28
2.	41%			
3.	53%			
4.	100%			
5.	48%			
6.	32%			
7.	81%			
8.	40%			
9.	7%			
10.	95%			
11.	82%			
12.	75%			
13.	42%			
14.	250%			
15.	125%			
16.	$33\frac{1}{3}$%			
17.	145%			
18.	$12\frac{1}{2}$%			
19.	112%			
20.	43%			

Fraction to decimal and percent

Name _____

Date _____

Rewrite the fractions in three ways.

	fraction (denominator of 100)	fraction (lowest terms)	decimal form	percent form
1.	$\frac{25}{100}$	$\frac{1}{4}$	0.25	25%
2.	$\frac{40}{100}$			
3.	$\frac{10}{100}$			
4.	$\frac{130}{100}$			
5.	$\frac{85}{100}$			
6.	$\frac{125}{100}$			
7.	$\frac{80}{100}$			
8.	$\frac{180}{100}$			
9.	$\frac{185}{100}$			
10.	$\frac{4}{100}$			
11.	$\frac{36}{100}$			
12.	$\frac{104}{100}$			
13.	$\frac{48}{100}$			
14.	$\frac{230}{100}$			
15.	$\frac{124}{100}$			
16.	$\frac{64}{100}$			
17.	$\frac{15}{100}$			
18.	$\frac{275}{100}$			
19.	$\frac{115}{100}$			
20.	$\frac{245}{100}$			

Fraction of a number

Percent of a number

Complete each of the following.

1. $\frac{1}{2}$ of 42 = *21*

2. $\frac{1}{2}$ of 84 = _____

3. $\frac{1}{2}$ of 150 = _____

4. $\frac{1}{2}$ of 28 = _____

5. $\frac{1}{2}$ of 114 = _____

6. $\frac{1}{4}$ of 36 = _____

7. $\frac{1}{4}$ of 48 = _____

8. $\frac{1}{4}$ of 112 = _____

9. $\frac{1}{4}$ of 216 = _____

10. $\frac{1}{4}$ of 200 = _____

11. $\frac{3}{5}$ of 85 = _____

12. $\frac{3}{5}$ of 150 = _____

13. $\frac{3}{5}$ of 55 = _____

14. $\frac{3}{5}$ of 105 = _____

15. $\frac{3}{5}$ of 65 = _____

16. 50% of 42 = _____

17. 50% of 84 = _____

18. 50% of 150 = _____

19. 50% of 28 = _____

20. 50% of 114 = _____

21. 25% of 36 = _____

22. 25% of 48 = _____

23. 25% of 112 = _____

24. 25% of 216 = _____

25. 25% of 200 = _____

26. 60% of 85 = _____

27. 60% of 150 = _____

28. 60% of 55 = _____

29. 60% of 105 = _____

30. 60% of 65 = _____

11

Fraction of a number
Finding percentage

Name _____

Date _____

Complete each of the following.

1. $\frac{7}{5}$ of 50 = _70_

2. $\frac{7}{5}$ of 120 = _____

3. $\frac{7}{5}$ of 115 = _____

4. $\frac{7}{5}$ of 180 = _____

5. $\frac{7}{5}$ of 205 = _____

6. 140% of 50 = _____

7. 140% of 120 = _____

8. 140% of 115 = _____

9. 140% of 180 = _____

10. 140% of 205 = _____

Find the percentages.

11. 36% of 13 = _4.68_

12. 51% of 28 = _____

13. 100% of 55 = _____

14. 16% of 56 = _____

15. 120% of 71 = _____

16. 7% of 432 = _____

17. 13% of 27 = _____

18. 27% of 92 = _____

19. 9% of 671 = _____

20. 101% of 65 = _____

Fraction of a number
Percent of a number

Name _____

Date _____

Complete each of the following.

1. $\frac{1}{2}$ of 24 = _12_

2. $\frac{1}{2}$ of 92 = _____

3. $\frac{1}{2}$ of 144 = _____

4. $\frac{1}{2}$ of 70 = _____

5. $\frac{1}{2}$ of 68 = _____

6. $\frac{1}{3}$ of 24 = _____

7. $\frac{1}{3}$ of 48 = _____

8. $\frac{1}{3}$ of 144 = _____

9. $\frac{1}{3}$ of 204 = _____

10. $\frac{1}{3}$ of 183 = _____

11. $\frac{4}{5}$ of 15 = _____

12. $\frac{4}{5}$ of 95 = _____

13. $\frac{4}{5}$ of 25 = _____

14. $\frac{4}{5}$ of 150 = _____

15. $\frac{4}{5}$ of 200 = _____

16. 50% of 24 = _____

17. 50% of 92 = _____

18. 50% of 144 = _____

19. 50% of 70 = _____

20. 50% of 68 = _____

21. $33\frac{1}{3}$% of 24 = _____

22. $33\frac{1}{3}$% of 48 = _____

23. $33\frac{1}{3}$% of 144 = _____

24. $33\frac{1}{3}$% of 204 = _____

25. $33\frac{1}{3}$% of 183 = _____

26. 80% of 15 = _____

27. 80% of 95 = _____

28. 80% of 25 = _____

29. 80% of 150 = _____

30. 80% of 200 = _____

Fraction of a number
Finding percentage

Name _____

Date _____

Complete each of the following.

1. $\frac{9}{4}$ of 16 = __36__

2. $\frac{9}{4}$ of 44 = _____

3. $\frac{9}{4}$ of 116 = _____

4. $\frac{9}{4}$ of 80 = _____

5. $\frac{9}{4}$ of 100 = _____

6. 225% of 16 = _____

7. 225% of 44 = _____

8. 225% of 116 = _____

9. 225% of 80 = _____

10. 225% of 100 = _____

Find the percentages.

11. 52% of 65 = __33.8__

12. 142% of 127 = _____

13. 92% of 143 = _____

14. 51% of 89 = _____

15. 27% of 415 = _____

16. 23% of 96 = _____

17. 78% of 32 = _____

18. 81% of 116 = _____

19. 143% of 38 = _____

20. 32% of 345 = _____

Fraction of a number
Percent of a number

Name _____

Date _____

Complete each of the following.

1. $\frac{3}{4}$ of 28 = _21_

2. $\frac{3}{4}$ of 80 = _____

3. $\frac{3}{4}$ of 224 = _____

4. $\frac{3}{4}$ of 172 = _____

5. $\frac{3}{4}$ of 364 = _____

6. $\frac{1}{3}$ of 39 = _____

7. $\frac{1}{3}$ of 144 = _____

8. $\frac{1}{3}$ of 99 = _____

9. $\frac{1}{3}$ of 282 = _____

10. $\frac{1}{3}$ of 384 = _____

11. $\frac{3}{8}$ of 72 = _____

12. $\frac{3}{8}$ of 96 = _____

13. $\frac{3}{8}$ of 104 = _____

14. $\frac{3}{8}$ of 32 = _____

15. $\frac{3}{8}$ of 184 = _____

16. 75% of 28 = _____

17. 75% of 80 = _____

18. 75% of 224 = _____

19. 75% of 172 = _____

20. 75% of 364 = _____

21. $33\frac{1}{3}$% of 39 = _____

22. $33\frac{1}{3}$% of 144 = _____

23. $33\frac{1}{3}$% of 99 = _____

24. $33\frac{1}{3}$% of 282 = _____

25. $33\frac{1}{3}$% of 384 = _____

26. $37\frac{1}{2}$% of 72 = _____

27. $37\frac{1}{2}$% of 96 = _____

28. $37\frac{1}{2}$% of 104 = _____

29. $37\frac{1}{2}$% of 32 = _____

30. $37\frac{1}{2}$% of 184 = _____

Fraction of a number   Name _____

Finding percentage   Date _____

_____

Complete each of the following.

1. $\frac{17}{10}$ of 80 = _/ 3 6_

6. 170% of 80 = _____

2. $\frac{17}{10}$ of 40 = _____

7. 170% of 40 = _____

3. $\frac{17}{10}$ of 200 = _____

8. 170% of 200 = _____

4. $\frac{17}{10}$ of 190 = _____

9. 170% of 190 = _____

5. $\frac{17}{10}$ of 240 = _____

10. 170% of 240 = _____

Find the percentages.

11. 37% of 92 = _34.04_

16. 52% of 117 = _____

12. 93% of 17 = _____

17. 78% of 62 = _____

13. 24% of 83 = _____

18. 143% of 22 = _____

14. 205% of 55 = _____

19. 76% of 108 = _____

15. 98% of 67 = _____

20. 43% of 39 = _____

Fraction of a number
Percent of a number

Name _____

Date _____

Complete each of the following.

1. $\frac{1}{8}$ of 80 = __10__

2. $\frac{1}{8}$ of 32 = _____

3. $\frac{1}{8}$ of 144 = _____

4. $\frac{1}{8}$ of 288 = _____

5. $\frac{1}{8}$ of 232 = _____

6. $\frac{1}{2}$ of 98 = _____

7. $\frac{1}{2}$ of 42 = _____

8. $\frac{1}{2}$ of 178 = _____

9. $\frac{1}{2}$ of 214 = _____

10. $\frac{1}{2}$ of 308 = _____

11. $\frac{7}{10}$ of 100 = _____

12. $\frac{7}{10}$ of 270 = _____

13. $\frac{7}{10}$ of 350 = _____

14. $\frac{7}{10}$ of 420 = _____

15. $\frac{7}{10}$ of 530 = _____

16. $12\frac{1}{2}$% of 80 = _____

17. $12\frac{1}{2}$% of 32 = _____

18. $12\frac{1}{2}$% of 144 = _____

19. $12\frac{1}{2}$% of 288 = _____

20. $12\frac{1}{2}$% of 232 = _____

21. 50% of 98 = _____

22. 50% of 42 = _____

23. 50% of 178 = _____

24. 50% of 214 = _____

25. 50% of 308 = _____

26. 70% of 100 = _____

27. 70% of 270 = _____

28. 70% of 350 = _____

29. 70% of 420 = _____

30. 70% of 530 = _____

Fraction of a number
Finding percentage

Name _____

Date _____

Complete each of the following.

1. $\frac{9}{5}$ of 25 = _45_

2. $\frac{9}{5}$ of 40 = _____

3. $\frac{9}{5}$ of 110 = _____

4. $\frac{9}{5}$ of 75 = _____

5. $\frac{9}{5}$ of 150 = _____

6. 180% of 25 = _____

7. 180% of 40 = _____

8. 180% of 110 = _____

9. 180% of 75 = _____

10. 180% of 150 = _____

Find the percentages.

11. 72% of 44 = _31.68_

12. 57% of 53 = _____

13. 44% of 72 = _____

14. 127% of 38 = _____

15. 68% of 68 = _____

16. 81% of 68 = _____

17. 62% of 38 = _____

18. 29% of 86 = _____

19. 57% of 36 = _____

20. 38% of 116 = _____

Fraction of a number
Percent of a number

Name _____

Date _____

Complete each of the following.

1. $\frac{4}{5}$ of 35 = _28_

2. $\frac{4}{5}$ of 75 = _____

3. $\frac{4}{5}$ of 60 = _____

4. $\frac{4}{5}$ of 105 = _____

5. $\frac{4}{5}$ of 225 = _____

6. $\frac{5}{8}$ of 64 = _____

7. $\frac{5}{8}$ of 80 = _____

8. $\frac{5}{8}$ of 120 = _____

9. $\frac{5}{8}$ of 72 = _____

10. $\frac{5}{8}$ of 248 = _____

11. $\frac{9}{10}$ of 100 = _____

12. $\frac{9}{10}$ of 50 = _____

13. $\frac{9}{10}$ of 140 = _____

14. $\frac{9}{10}$ of 230 = _____

15. $\frac{9}{10}$ of 350 = _____

16. 80% of 35 = _____

17. 80% of 75 = _____

18. 80% of 60 = _____

19. 80% of 105 = _____

20. 80% of 225 = _____

21. $62\frac{1}{2}$% of 64 = _____

22. $62\frac{1}{2}$% of 80 = _____

23. $62\frac{1}{2}$% of 120 = _____

24. $62\frac{1}{2}$% of 72 = _____

25. $62\frac{1}{2}$% of 248 = _____

26. 90% of 100 = _____

27. 90% of 50 = _____

28. 90% of 140 = _____

29. 90% of 230 = _____

30. 90% of 350 = _____

Fraction of a number
Finding percentage

Name _____

Date _____

Complete each of the following.

1. $\frac{7}{4}$ of 16 = _28_

2. $\frac{7}{4}$ of 72 = _____

3. $\frac{7}{4}$ of 148 = _____

4. $\frac{7}{4}$ of 116 = _____

5. $\frac{7}{4}$ of 232 = _____

6. 175% of 16 = _____

7. 175% of 72 = _____

8. 175% of 148 = _____

9. 175% of 116 = _____

10. 175% of 232 = _____

Find the percentages.

11. 32% of 61 = _19.52_

12. 53% of 78 = _____

13. 126% of 37 = _____

14. 74% of 21 = _____

15. 146% of 37 = _____

16. 15% of 198 = _____

17. 73% of 28 = _____

18. 44% of 92 = _____

19. 23% of 192 = _____

20. 184% of 28 = _____

Fraction to percent
Finding percentage

Name _____

Date _____

Complete each of the following.

1. $\frac{1}{2} = $ __50__ %   6. $\frac{1}{5} = $ _____ %   11. $\frac{1}{8} = $ _____ %   16. $\frac{7}{10} = $ _____ %

2. $\frac{1}{3} = $ _____ %   7. $\frac{2}{5} = $ _____ %   12. $\frac{3}{8} = $ _____ %   17. $\frac{3}{10} = $ _____ %

3. $\frac{2}{3} = $ _____ %   8. $\frac{3}{5} = $ _____ %   13. $\frac{5}{8} = $ _____ %   18. $\frac{9}{10} = $ _____ %

4. $\frac{1}{4} = $ _____ %   9. $\frac{4}{5} = $ _____ %   14. $\frac{7}{8} = $ _____ %   19. $\frac{5}{4} = $ _____ %

5. $\frac{3}{4} = $ _____ %   10. $\frac{7}{5} = $ _____ %   15. $\frac{9}{8} = $ _____ %   20. $\frac{11}{8} = $ _____ %

21. 50% of 20 = __10__   29. 125% of 72 = _____

22. 75% of 40 = _____   30. 40% of 55 = _____

23. 140% of 35 = _____   31. 75% of 32 = _____

24. $37\frac{1}{2}$% of 96 = _____   32. $33\frac{1}{3}$% of 60 = _____

25. 30% of 200 = _____   33. 90% of 120 = _____

26. 25% of 20 = _____   34. $137\frac{1}{2}$% of 112 = _____

27. 60% of 15 = _____   35. $87\frac{1}{2}$% of 56 = _____

28. $12\frac{1}{2}$% of 64 = _____   36. $166\frac{2}{3}$% of 321 = _____

21

Percent to decimal
Finding percentage

Name _____

Date _____

Rewrite each of the following in decimal form.

1. 48% = _0.48_   6. 17% = _____   11. 167% = _____   16. 250% = _____

2. 92% = _____   7. 87% = _____   12. 105% = _____   17. 700% = _____

3. 56% = _____   8. 42% = _____   13. 110% = _____   18. 1% = _____

4. 12% = _____   9. 31% = _____   14. 0.7% = _____   19. 17.5% = _____

5. 59% = _____   10. 125% = _____   15. 3% = _____   20. 9.3% = _____

Complete each of the following.

21. 48% of 29 = _13.92_   29. 125% of 92 = _____

22. 92% of 76 = _____   30. 250% of 13 = _____

23. 12% of 93 = _____   31. 167% of 42 = _____

24. 59% of 87 = _____   32. 105% of 27 = _____

25. 105% of 46 = _____   33. 0.7% of 324 = _____

26. 17% of 35 = _____   34. 1% of 512 = _____

27. 87% of 43 = _____   35. 17.5% of 22 = _____

28. 31% of 75 = _____   36. 2.8% of 315 = _____

Fraction to percent
Finding percentage

Name _____

Date _____

Complete each of the following.

1. $\frac{5}{8}$ = __62½__ %   6. $\frac{2}{3}$ = _____ %   11. $\frac{3}{4}$ = _____ %   16. $\frac{9}{5}$ = _____ %

2. $\frac{1}{3}$ = _____ %   7. $\frac{5}{4}$ = _____ %   12. $\frac{7}{8}$ = _____ %   17. $\frac{9}{10}$ = _____ %

3. $\frac{1}{4}$ = _____ %   8. $\frac{3}{8}$ = _____ %   13. $\frac{13}{8}$ = _____ %   18. $\frac{3}{5}$ = _____ %

4. $\frac{11}{8}$ = _____ %   9. $\frac{15}{4}$ = _____ %   14. $\frac{7}{10}$ = _____ %   19. $\frac{5}{3}$ = _____ %

5. $\frac{4}{5}$ = _____ %   10. $\frac{17}{5}$ = _____ %   15. $\frac{2}{9}$ = _____ %   20. $\frac{3}{10}$ = _____ %

21. 40% of 95 = __38__   29. 62½% of 80 = _____

22. 20% of 425 = _____   30. 25% of 224 = _____

23. 37½% of 216 = _____   31. 70% of 320 = _____

24. 340% of 75 = _____   32. 90% of 110 = _____

25. 140% of 105 = _____   33. 166⅔% of 63 = _____

26. 25% of 196 = _____   34. 75% of 236 = _____

27. 80% of 265 = _____   35. 66⅔% of 108 = _____

28. 22⅔% of 81 = _____   36. 87½% of 184 = _____

Percent to decimal
Finding percentage

Name _____

Date _____

Rewrite each of the following in decimal form.

1. 36% = _0.36_   6. 31% = _____   11. 27% = _____   16. 8.2% = _____

2. 57% = _____   7. 19% = _____   12. 5.4% = _____   17. 17.4% = _____

3. 46% = _____   8. 16% = _____   13. 3.7% = _____   18. 13.6% = _____

4. 8% = _____   9. 13% = _____   14. 8.1% = _____   19. 15.9% = _____

5. 13% = _____   10. 132% = _____   15. 16.3% = _____   20. 28.1% = _____

Complete each of the following.

21. 36% of 43 = _15.48_   29. 8.1% of 26 = _____

22. 57% of 28 = _____   30. 3.7% of 48 = _____

23. 46% of 32 = _____   31. 8.2% of 27 = _____

24. 13% of 143 = _____   32. 5.4% of 63 = _____

25. 16% of 86 = _____   33. 13.6% of 67 = _____

26. 31% of 29 = _____   34. 15.9% of 54 = _____

27. 19% of 75 = _____   35. 28.1% of 92 = _____

28. 132% of 49 = _____   36. 1.8% of 110 = _____

Fraction to percent
Finding percentage

Complete each of the following.

1. $\frac{3}{8}$ = _37½_ %   6. $\frac{3}{5}$ = _____ %   11. $\frac{11}{8}$ = _____ % 16. $\frac{1}{5}$ = _____ %

2. $\frac{5}{4}$ = _____ %   7. $\frac{7}{8}$ = _____ %   12. $\frac{1}{2}$ = _____ %   17. $\frac{3}{10}$ = _____ %

3. $\frac{8}{5}$ = _____ %   8. $\frac{1}{4}$ = _____ %   13. $\frac{2}{5}$ = _____ %   18. $\frac{5}{3}$ = _____ %

4. $\frac{1}{3}$ = _____ %   9. $\frac{2}{3}$ = _____ %   14. $\frac{5}{8}$ = _____ %   19. $\frac{3}{4}$ = _____ %

5. $\frac{13}{4}$ = _____ % 10. $\frac{7}{10}$ = _____ % 15. $\frac{4}{5}$ = _____ %   20. $\frac{5}{9}$ = _____ %

21. 70% of 130 = _91_   29. 175% of 56 = _____

22. $137\frac{1}{2}$% of 40 = _____   30. 80% of 135 = _____

23. 50% of 162 = _____   31. $33\frac{1}{3}$% of 93 = _____

24. 40% of 95 = _____   32. 125% of 64 = _____

25. $62\frac{1}{2}$% of 96 = _____   33. 160% of 105 = _____

26. $87\frac{1}{2}$% of 128 = _____   34. 60% of 125 = _____

27. 20% of 40 = _____   35. $166\frac{2}{3}$% of 81 = _____

28. $37\frac{1}{2}$ % of 88 = _____   36. $12\frac{1}{2}$% of 328 = _____

25

Name _____

Date _____

Rewrite each of the following in decimal form.

1. 26% = _0.26_   6. 12% = _____   11. 6.9% = _____   16. 3.2% = _____

2. 5% = _____   7. 29% = _____   12. 7.62% = _____   17. 5.14% = _____

3. 16% = _____   8. 32% = _____   13. 0.31% = _____   18. 0.23% = _____

4. 82% = _____   9. 9% = _____   14. 810% = _____   19. 0.3% = _____

5. 8% = _____   10. 17% = _____   15. 9.6% = _____   20. 114% = _____

Complete each of the following.

21. 26% of 87 = _22.62_   29. 9% of 243 = _____

22. 16% of 96 = _____   30. 17% of 106 = _____

23. 82% of 16 = _____   31. 3.2% of 69 = _____

24. 8% of 503 = _____   32. 5.14% of 73 = _____

25. 810% of 43 = _____   33. 0.23% of 53 = _____

26. 12% of 67 = _____   34. 0.3% of 2731 = _____

27. 29% of 56 = _____   35. 114% of 29 = _____

28. 32% of 82 = _____   36. 0.03% of 804 = _____

Fraction to percent
Finding percentage

Name _____

Date _____

Complete each of the following.

1. $\frac{9}{10}$ = __*90*__ %   6. $\frac{9}{4}$ = _____ %   11. $\frac{2}{5}$ = _____ %   16. $\frac{7}{5}$ = _____ %

2. $\frac{1}{8}$ = _____ %   7. $\frac{1}{4}$ = _____ %   12. $\frac{17}{8}$ = _____ %   17. $\frac{15}{8}$ = _____ %

3. $\frac{13}{10}$ = _____ %   8. $\frac{3}{4}$ = _____ %   13. $\frac{5}{8}$ = _____ %   18. $\frac{1}{2}$ = _____ %

4. $\frac{5}{4}$ = _____ %   9. $\frac{3}{8}$ = _____ %   14. $\frac{13}{4}$ = _____ %   19. $\frac{9}{5}$ = _____ %

5. $\frac{1}{9}$ = _____ %   10. $\frac{7}{8}$ = _____ %   15. $\frac{7}{10}$ = _____ %   20. $\frac{11}{9}$ = _____ %

21. $37\frac{1}{2}$% of 16 = __*6*__   29. $12\frac{1}{2}$% of 88 = _____

22. 75% of 48 = _____   30. 25% of 188 = _____

23. 70% of 30 = _____   31. 75% of 204 = _____

24. $87\frac{1}{2}$% of 32 = _____   32. 225% of 92 = _____

25. 50% of 150 = _____   33. 130% of 150 = _____

26. $122\frac{2}{9}$% of 18 = _____   34. $11\frac{1}{9}$% of 63 = _____

27. 90% of 140 = _____   35. 125% of 124 = _____

28. 180% of 65 = _____   36. $166\frac{2}{3}$% of 321 = _____

Copyright © 1981 by Dale Seymour Publications.

27

Rewrite each of the following in decimal form.

1. 22% = _0.22_   **6.** 69% = _____   **11.** 17.2% = _____   **16.** 9.2% = _____

2. 13% = _____   **7.** 52% = _____   **12.** 18.1% = _____   **17.** 0.35% = _____

3. 72% = _____   **8.** 55% = _____   **13.** 5.3% = _____   **18.** 1.4% = _____

4. 81% = _____   **9.** 32% = _____   **14.** 113% = _____   **19.** 9.5% = _____

5. 24% = _____   **10.** 137% = _____   **15.** 15.1% = _____   **20.** 8.1% = _____

Complete each of the following.

21. 22% of 81 = _17.82_   **29.** 18.1% of 15 = _____

22. 13% of 65 = _____   **30.** 0.3% of 98 = _____

23. 52% of 19 = _____   **31.** 15.1% of 76 = _____

24. 81% of 63 = _____   **32.** 9.2% of 57 = _____

25. 24% of 60 = _____   **33.** 0.35% of 600 = _____

26. 72% of 17 = _____   **34.** 1.4% of 28 = _____

27. 32% of 63 = _____   **35.** 113% of 56 = _____

28. 17.2% of 32 = _____   **36.** 17.3% of 67 = _____

Fraction to percent
Finding percentage

Name _____

Date _____

Complete each of the following.

1. $\frac{2}{5}$ = __40__ %   6. $\frac{5}{3}$ = _____ %   11. $\frac{4}{5}$ = _____ %   16. $\frac{3}{10}$ = _____%

2. $\frac{2}{3}$ = _____ %   7. $\frac{3}{8}$ = _____ %   12. $\frac{1}{3}$ = _____ %   17. $\frac{7}{4}$ = _____ %

3. $\frac{5}{4}$ = _____ %   8. $\frac{3}{2}$ = _____ %   13. $\frac{1}{2}$ = _____ %   18. $\frac{7}{3}$ = _____ %

4. $\frac{7}{5}$ = _____ %   9. $\frac{3}{5}$ = _____ %   14. $\frac{1}{8}$ = _____ %   19. $\frac{1}{9}$ = _____ %

5. $\frac{7}{10}$ = _____%   10. $\frac{3}{4}$ = _____ %   15. $\frac{1}{5}$ = _____ %   20. $\frac{13}{10}$ = _____ %

21. 125% of 36 = __45__

22. 70% of 90 = _____

23. $166\frac{2}{3}$% of 81 = _____

24. 30% of 110 = _____

25. 130% of 50 = _____

26. 80% of 65 = _____

27. 40% of 125 = _____

28. 150% of 90 = _____

29. 140% of 85 = _____

30. $37\frac{1}{2}$% of 32 = _____

31. 75% of 108 = _____

32. 175% of 92 = _____

33. $66\frac{2}{3}$% of 54 = _____

34. $11\frac{1}{9}$% of 90 = _____

35. $33\frac{1}{3}$% of 66 = _____

36. 125% of 36 = _____

Percent to decimal
Finding percentage

Name _____

Date _____

Rewrite each of the following in decimal form.

1. 42% = _0.42_   6. 83% = _____   11. 12.2% = _____   16. 3.5% = _____

2. 35% = _____   7. 9% = _____   12. 8.7% = _____   17. 12.9% = _____

3. 58% = _____   8. 72% = _____   13. 115% = _____   18. 5.2% = _____

4. 7% = _____   9. 77% = _____   14. 0.32% = _____   19. 14.1% = _____

5. 16% = _____   10. 89% = _____   15. 6.9% = _____   20. 0.03% = _____

Complete each of the following.

21. 42% of 94 = _39.48_

22. 35% of 27 = _____

23. 58% of 63 = _____

24. 77% of 217 = _____

25. 16% of 405 = _____

26. 9% of 376 = _____

27. 7% of 931 = _____

28. 12.2% of 76 = _____

29. 8.7% of 335 = _____

30. 115% of 27 = _____

31. 0.32% of 93 = _____

32. 6.9% of 102 = _____

33. $\frac{3}{5}$% of 61 = _____

34. 12.9% of 53 = _____

35. 14.1% of 57 = _____

36. 2.3% of 15 = _____

Percent of a number

Find the percentages.

1. 12% of 32 = _3.84_

2. 25% of 76 = _____

3. 13% of 93 = _____

4. 20% of 85 = _____

5. 37% of 83 = _____

6. 10% of 110 = _____

7. 6.1% of 121 = _____

8. 60% of 95 = _____

9. 12.3% of 69 = _____

10. 80% of 155 = _____

11. 75% of 128 = _____

12. 15.2% of 7.3 = _____

13. 30% of 150 = _____

14. 28.2% of 36 = _____

15. 50% of 86 = _____

16. 17.5% of 7.2 = _____

17. 90% of 380 = _____

18. 63.2% of 99 = _____

19. 40% of 275 = _____

20. 8.7% of 68 = _____

Finding percent

Name _____

Date _____

Find the percents.

1. $\frac{1}{2}$ is what percent? _50%_

2. $\frac{3}{5}$ is what percent? _____

3. $\frac{1}{4}$ is what percent? _____

4. $\frac{1}{10}$ is what percent? _____

5. $\frac{4}{5}$ is what percent? _____

6. $\frac{3}{4}$ is what percent? _____

7. $\frac{3}{10}$ is what percent? _____

8. $\frac{5}{8}$ is what percent? _____

9. $\frac{2}{5}$ is what percent? _____

10. $\frac{7}{8}$ is what percent? _____

11. 4 is what percent of 5? _____

12. 7 is what percent of 10? _____

13. 1 is what percent of 2? _____

14. 3 is what percent of 4? _____

15. 5 is what percent of 8? _____

16. 2 is what percent of 5? _____

17. 3 is what percent of 10? _____

18. 1 is what percent of 4? _____

19. 3 is what percent of 5? _____

20. 6 is what percent of 12? _____

Percent of a number

Find the percentages.

1. 17% of 28 = _4.76_

2. 8% of 627 = _____

3. 10% of 230 = _____

4. 14.3% of 97 = _____

5. 5.6% of 32 = _____

6. 25% of 112 = _____

7. 9.9% of 76 = _____

8. 12% of 322 = _____

9. 50% of 144 = _____

10. 6% of 1365 = _____

11. 75% of 384 = _____

12. 19.3% of 63 = _____

13. 27% of 115 = _____

14. 50% of 562 = _____

15. 16% of 76 = _____

16. 39% of 58 = _____

17. 80% of 325 = _____

18. 67% of 83 = _____

19. 53.1% of 93 = _____

20. 17.4% of 67 = _____

Name _____

Date _____

Find the percents.

1. $\frac{1}{3}$ is what percent? $33\frac{1}{3}\%$

2. $\frac{3}{4}$ is what percent? _____

3. $\frac{3}{8}$ is what percent? _____

4. $\frac{3}{5}$ is what percent? _____

5. $\frac{1}{2}$ is what percent? _____

6. $\frac{4}{5}$ is what percent? _____

7. $\frac{2}{3}$ is what percent? _____

8. $\frac{5}{8}$ is what percent? _____

9. $\frac{1}{4}$ is what percent? _____

10. $\frac{7}{10}$ is what percent? _____

11. 3 is what percent of 5? _____

12. 5 is what percent of 8? _____

13. 9 is what percent of 10? _____

14. 1 is what percent of 2? _____

15. 1 is what percent of 4? _____

16. 7 is what percent of 10? _____

17. 3 is what percent of 4? _____

18. 3 is what percent of 8? _____

19. 3 is what percent of 10? _____

20. 2 is what percent of 5? _____

Percent of a number

Find the percentages.

1. 13% of 861 = _111.93_

2. 50% of 32876 = _____

3. 27% of 693 = _____

4. 10% of 630 = _____

5. 123% of 76 = _____

6. 90% of 980 = _____

7. 42% of 93 = _____

8. 80% of 435 = _____

9. 7.6% of 32 = _____

10. 30% of 5630 = _____

11. 8.3% of 67 = _____

12. 25% of 396 = _____

13. 112% of 13 = _____

14. 12.5% of 64 = _____

15. 56% of 163 = _____

16. 75% of 444 = _____

17. 42% of 94 = _____

18. 70% of 1220 = _____

19. 60% of 555 = _____

20. 76% of 23 = _____

Name _____

Date _____

Find the percents.

1. $\frac{9}{10}$ is what percent? _90%_

2. $\frac{5}{8}$ is what percent? _____

3. $\frac{2}{3}$ is what percent? _____

4. $\frac{3}{5}$ is what percent? _____

5. $\frac{1}{2}$ is what percent? _____

6. $\frac{1}{3}$ is what percent? _____

7. $\frac{3}{8}$ is what percent? _____

8. $\frac{3}{4}$ is what percent? _____

9. $\frac{3}{20}$ is what percent? _____

10. $\frac{7}{12}$ is what percent? _____

11. 8 is what percent of 10? _____

12. 4 is what percent of 10? _____

13. 9 is what percent of 10? _____

14. 11 is what percent of 10? _____

15. 5 is what percent of 10? _____

16. 15 is what percent of 10? _____

17. 6 is what percent of 10? _____

18. 20 is what percent of 10? _____

19. 3 is what percent of 10? _____

20. 2 is what percent of 10? _____

Percent of a number

Name _____

Date _____

Find the percentages.

1. 13.2% of 83 = _10.956_

2. 10% of 830 = _____

3. 15% of 1640 = _____

4. 50% of 1550 = _____

5. 7% of 1382 = _____

6. 23.1% of 680 = _____

7. 126% of 32 = _____

8. 75% of 3004 = _____

9. 59% of 692 = _____

10. 30% of 670 = _____

11. $66\frac{2}{3}$% of 69 = _____

12. 25% of 3984 = _____

13. 7.8% of 64 = _____

14. 60% of 375 = _____

15. 12.5% of 96 = _____

16. 23.3% of 11 = _____

17. 80% of 625 = _____

18. 66% of 58 = _____

19. 40% of 835 = _____

20. $33\frac{1}{3}$% of 54 = _____

Finding percent

Find the percents.

1. $\frac{3}{5}$ is what percent? _60%_

2. $\frac{3}{10}$ is what percent? _____

3. $\frac{3}{4}$ is what percent? _____

4. $\frac{1}{3}$ is what percent? _____

5. $\frac{7}{8}$ is what percent? _____

6. $\frac{5}{8}$ is what percent? _____

7. $\frac{2}{7}$ is what percent? _____

8. $\frac{7}{10}$ is what percent? _____

9. $\frac{2}{3}$ is what percent? _____

10. $\frac{9}{10}$ is what percent? _____

11. 7 is what percent of 14? _____

12. 9 is what percent of 3? _____

13. 12 is what percent of 8? _____

14. 6 is what percent of 5? _____

15. 3 is what percent of 4? _____

16. 16 is what percent of 20? _____

17. 1 is what percent of 3? _____

18. 5 is what percent of 15? _____

19. 7 is what percent of 10? _____

20. 5 is what percent of 8? _____

38

Percent of a number

Find the percentages.

1. 27% of 133 = _35.91_

2. 75% of 1324 = _____

3. 16.2% of 41 = _____

4. 10% of 46320 = _____

5. 123% of 72 = _____

6. 25% of 1684 = _____

7. 4.3% of 67 = _____

8. 5.4% of 32 = _____

9. 60% of 905 = _____

10. 67% of 302 = _____

11. 80% of 665 = _____

12. 65% of 38 = _____

13. $33\frac{1}{3}$% of 99 = _____

14. $12\frac{1}{2}$% of 128 = _____

15. 54% of 381 = _____

16. $66\frac{2}{3}$% of 333 = _____

17. 77% of 62 = _____

18. 50% of 1000 = _____

19. 38.1% of 53 = _____

20. 70% of 830 = _____

Name _____

Date _____

Find the percents.

1. $\frac{3}{5}$ is what percent? _60%_

2. $\frac{1}{3}$ is what percent? _____

3. $\frac{1}{2}$ is what percent? _____

4. $\frac{5}{8}$ is what percent? _____

5. $\frac{4}{5}$ is what percent? _____

6. $\frac{1}{4}$ is what percent? _____

7. $\frac{2}{3}$ is what percent? _____

8. $\frac{3}{8}$ is what percent? _____

9. $\frac{7}{10}$ is what percent? _____

10. $\frac{7}{20}$ is what percent? _____

11. 9 is what percent of 18? _____

12. 9 is what percent of 27? _____

13. 9 is what percent of 36? _____

14. 4 is what percent of 16? _____

15. 8 is what percent of 16? _____

16. 12 is what percent of 16? _____

17. 3 is what percent of 10? _____

18. 15 is what percent of 10? _____

19. 12 is what percent of 18? _____

20. 11 is what percent of 22? _____

Sales

Name _____

Date _____

When you buy an item on sale or at a discount price, it is helpful to know how much money you save. For example, suppose a store gives a 10% discount on a $9.00 record. You figure the savings as follows.

10% of $9.00 = 0.10 × $9.00 = $0.90
The sale price is $9.00 − $0.90 or $8.10.

Find the amount saved and the sale price.

Item: record
Regular Price: $9.00
Discount: 10%
Savings: $0.90
Sale Price: $8.10

1. Item: cassette tape player
Regular Price: $38.90
Discount: 10%
Savings: _____
Sale Price: _____

4. Item: hair dryer
Regular Price: $28.50
Discount: 40%
Savings: _____
Sale Price: _____

2. Item: instamatic camera
Regular Price: $35.45
Discount: 40%
Savings: _____
Sale Price: _____

5. Item: calculator
Regular Price: $19.75
Discount: 30%
Savings: _____
Sale Price: _____

3. Item: digital watch
Regular Price: $21.96
Discount: 25%
Savings: _____
Sale Price: _____

6. Item: pen/pencil set
Regular Price: $27.50
Discount: 40%
Savings: _____
Sale Price: _____

Name _____

Date _____

Complete the following table.

	item	regular price	discount	savings	sale price
1.	1 pair slacks	$18.25	20%	$3.65	$14.60
2.	1 blouse	$16.75	$33\frac{1}{3}\%$		
3.	1 swimsuit	$24.50	40%		
4.	1 shirt	$18.30	10%		
5.	1 pair jeans	$16.95	20%		
6.	1 pair shoes	$16.40	$12\frac{1}{2}\%$		
7.	1 pair shorts	$14.30	15%		
8.	1 belt	$ 7.90	10%		

Sales

Name _____

Date _____

When you buy an item on sale or at a discount price, it is helpful to know how much money you save. For example, suppose a store gives a 10% discount on a $9.00 record. You figure the savings as follows.

Item: record
Regular Price: $9.00
Discount: 10%
Savings: $0.90
Sale Price: $8.10

10% of $9.00 = 0.10 × $9.00 = $0.90
The sale price is $9.00 − $0.90 or $8.10.

Find the amount saved and the sale price.

1. Item: cassette tape player Regular Price: $45.40 Discount: 25% Savings: _____ Sale Price: _____	**4.** Item: hair dryer Regular Price: $21.50 Discount: 20% Savings: _____ Sale Price: _____
2. Item: instamatic camera Regular Price: $21.50 Discount: 30% Savings: _____ Sale Price: _____	**5.** Item: calculator Regular Price: $26.00 Discount: 15% Savings: _____ Sale Price: _____
3. Item: digital watch Regular Price: $28.50 Discount: 20% Savings: _____ Sale Price: _____	**6.** Item: pen/pencil set Regular Price: $27.00 Discount: 30% Savings: _____ Sale Price: _____

Copyright © 1981 by Dale Seymour Publications.

43

Savings and sale price

Name _____

Date _____

Complete the following table.

	item	regular price	discount	savings	sale price
1.	1 pair slacks	$15.00	$33\frac{1}{3}\%$	$5.00	$10.00
2.	1 blouse	$19.48	25%		
3.	1 swimsuit	$16.00	$12\frac{1}{2}\%$		
4.	1 shirt	$12.75	40%		
5.	1 pair jeans	$16.50	20%		
6.	1 pair shoes	$16.10	10%		
7.	1 pair shorts	$12.00	$33\frac{1}{3}\%$		
8.	1 belt	$ 8.00	$12\frac{1}{2}\%$		

Sales

Name _____

Date _____

When you buy an item on sale or at a discount price, it is helpful to know how much money you save. For example, suppose a store gives a 10% discount on a $9.00 record. You figure the savings as follows.

10% of $9.00 = 0.10 × $9.00 = $0.90
The sale price is $9.00 − $0.90 or $8.10.

Find the amount saved and the sale price.

Item: record
Regular Price: $9.00
Discount: 10%
Savings: $0.90
Sale Price: $8.10

---

1. Item: cassette tape player
Regular Price: $59.95
Discount: 40%
Savings: _____
Sale Price: _____

4. Item: hair dryer
Regular Price: $18.80
Discount: 15%
Savings: _____
Sale Price: _____

2. Item: instamatic camera
Regular Price: $28.80
Discount: 25%
Savings: _____
Sale Price: _____

5. Item: calculator
Regular Price: $19.80
Discount: 10%
Savings: _____
Sale Price: _____

3. Item: digital watch
Regular Price: $25.90
Discount: 30%
Savings: _____
Sale Price: _____

6. Item: pen/pencil set
Regular Price: $32.00
Discount: 25%
Savings: _____
Sale Price: _____

Copyright © 1981 by Dale Seymour Publications.

45

Savings and sale price

Name _____

Date _____

Complete the following table.

	item	regular price	discount	savings	sale price
1.	1 pair slacks	$14.00	10%	$1.40	$12.60
2.	1 blouse	$26.50	30%		
3.	1 swimsuit	$24.00	$33\frac{1}{3}$%		
4.	1 shirt	$16.50	40%		
5.	1 pair jeans	$17.40	25%		
6.	1 pair shoes	$18.00	$33\frac{1}{3}$%		
7.	1 pair shorts	$12.95	20%		
8.	1 belt	$ 9.00	$33\frac{1}{3}$%		

46

Sales

Name _____

Date _____

When you buy an item on sale or at a discount price, it is helpful to know how much money you save. For example, suppose a store gives a 10% discount on a $9.00 record. You figure the savings as follows.

10% of $9.00 = 0.10 × $9.00 = $0.90
The sale price is $9.00 − $0.90 or $8.10.

Find the amount saved and the sale price.

Item: record
Regular Price: $9.00
Discount: 10%
Savings: $0.90
Sale Price: $8.10

1. Item: cassette tape player
   Regular Price: $39.95
   Discount: 20%
   Savings: _____
   Sale Price: _____

4. Item: hair dryer
   Regular Price: $23.44
   Discount: 25%
   Savings: _____
   Sale Price: _____

2. Item: instamatic camera
   Regular Price: $17.50
   Discount: 10%
   Savings: _____
   Sale Price: _____

5. Item: calculator
   Regular Price: $25.50
   Discount: 20%
   Savings: _____
   Sale Price: _____

3. Item: digital watch
   Regular Price: $32.40
   Discount: 25%
   Savings: _____
   Sale Price: _____

6. Item: pen/pencil set
   Regular Price: $26.30
   Discount: 20%
   Savings: _____
   Sale Price: _____

Savings and sale price

Name _____

Date _____

Complete the following table.

	item	regular price	discount	savings	sale price
1.	1 pair slacks	$23.95	40%	$9.58	$14.37
2.	1 blouse	$16.20	20%		
3.	1 swimsuit	$23.40	25%		
4.	1 shirt	$ 9.20	10%		
5.	1 pair jeans	$14.40	15%		
6.	1 pair shoes	$18.50	30%		
7.	1 pair shorts	$12.50	20%		
8.	1 belt	$ 8.95	20%		

Sales

Name _____

Date _____

When you buy an item on sale or at a discount price, it is helpful to know how much money you save. For example, suppose a store gives a 10% discount on a $9.00 record. You figure the savings as follows.

10% of $9.00 = 0.10 × $9.00 = $0.90
The sale price is $9.00 − $0.90 or $8.10.

Find the amount saved and the sale price.

Item: record
Regular Price: $9.00
Discount: 10%
Savings: $0.90
Sale Price: $8.10

1. Item: cassette tape player
   Regular Price: $49.95
   Discount: 20%
   Savings: _____
   Sale Price: _____

4. Item: hair dryer
   Regular Price: $21.50
   Discount: 10%
   Savings: _____
   Sale Price: _____

2. Item: instamatic camera
   Regular Price: $32.00
   Discount: 25%
   Savings: _____
   Sale Price: _____

5. Item: calculator
   Regular Price: $27.44
   Discount: 25%
   Savings: _____
   Sale Price: _____

3. Item: digital watch
   Regular Price: $28.50
   Discount: 10%
   Savings: _____
   Sale Price: _____

6. Item: pen/pencil set
   Regular price: $35.00
   Discount: 15%
   Savings: _____
   Sale Price: _____

Name _____

Date _____

Complete the following table.

	item	regular price	discount	savings	sale price
1.	1 pair slacks	$17.50	20%	$3.50	$14.00
2.	1 blouse	$18.40	25%		
3.	1 swimsuit	$22.00	20%		
4.	1 shirt	$10.95	40%		
5.	1 pair jeans	$18.60	15%		
6.	1 pair shoes	$16.20	20%		
7.	1 pair shorts	$14.40	25%		
8.	1 belt	$10.00	10%		

Fraction to percent

Name _____

Date _____

Complete each of the following.

1. $\frac{1}{2}$ is _50_ percent.

2. $\frac{2}{5}$ is _____ percent.

3. $\frac{1}{8}$ is _____ percent.

4. $\frac{1}{3}$ is _____ percent.

5. $\frac{4}{5}$ is _____ percent.

6. $\frac{5}{8}$ is _____ percent.

7. $\frac{2}{3}$ is _____ percent.

8. $\frac{3}{4}$ is _____ percent.

9. $\frac{3}{10}$ is _____ percent.

10. $\frac{3}{20}$ is _____ percent.

11. 8 is _____ percent of 16.

12. 3 is _____ percent of 5.

13. 2 is _____ percent of 10.

14. 4 is _____ percent of 16.

15. 9 is _____ percent of 18.

16. 7 is _____ percent of 21.

17. 3 is _____ percent of 4.

18. 5 is _____ percent of 25.

19. 11 is _____ percent of 44.

20. 12 is _____ percent of 36.

Finding the whole
Finding percentage

Name _____

Date _____

Complete each of the following.

1. 12 is 25% of _48_ .

2. 18 is 50% of _____ .

3. 15 is 30% of _____ .

4. 9 is 10% of _____ .

5. 14 is $12\frac{1}{2}$% of _____ .

6. 15 is $33\frac{1}{3}$% of _____ .

7. 38 is 40% of _____ .

8. 165 is 15% of _____ .

9. 90 is 75% of _____ .

10. 84 is 80% of _____ .

11. 23.36 is 73% of _____ .

12. 29.76 is 62% of _____ .

13. 53.55 is 51% of _____ .

14. 1.92 is 12% of _____ .

15. 7.2 is 8% of _____ .

16. 6.21 is 23% of _____ .

17. 6.12 is 17% of _____ .

18. 15.39 is 19% of _____ .

19. 8.32 is 16% of _____ .

20. 4.62 is 33% of _____ .

Find the percentages.

21. 28.2% of 17 = _4.794_

22. 15.7% of 65 = _____

23. $66\frac{2}{3}$% of 48 = _____

24. 19.4% of 62 = _____

25. 43.8% of 29 = _____

26. $37\frac{1}{2}$% of 24 = _____

Fraction to percent

Complete each of the following.

1. $\frac{1}{2}$ is _50_ percent.

2. $\frac{4}{5}$ is _____ percent.

3. $\frac{3}{4}$ is _____ percent.

4. $\frac{3}{10}$ is _____ percent.

5. $\frac{2}{5}$ is _____ percent.

6. $\frac{1}{4}$ is _____ percent.

7. $\frac{9}{10}$ is _____ percent.

8. $\frac{1}{8}$ is _____ percent.

9. $\frac{1}{5}$ is _____ percent.

10. $\frac{5}{8}$ is _____ percent.

11. 3 is _____ percent of 5.

12. 7 is _____ percent of 10.

13. 1 is _____ percent of 2.

14. 4 is _____ percent of 5.

15. 7 is _____ percent of 8.

16. 3 is _____ percent of 4.

17. 5 is _____ percent of 10.

18. 30 is _____ percent of 40.

19. 16 is _____ percent of 64.

20. 24 is _____ percent of 32.

Finding the whole
Finding percentage

Name _____

Date _____

Complete each of the following.

1. 4 is 50% of __8__.

2. 8 is 25% of _____.

3. 9 is 90% of _____.

4. 21 is 75% of _____.

5. 12 is $66\frac{2}{3}$% of _____.

6. 18 is 40% of _____.

7. 26 is 50% of _____.

8. 40 is $62\frac{1}{2}$% of _____.

9. 27 is 60% of _____.

10. 18 is $12\frac{1}{2}$% of _____.

11. 2.85 is 15% of _____.

12. 8.32 is 26% of _____.

13. 2.38 is 17% of _____.

14. 24.94 is 29% of _____.

15. 3.84 is 16% of _____.

16. 29.16 is 36% of _____.

17. 23.92 is 52% of _____.

18. 43.32 is 76% of _____.

19. 19.44 is 81% of _____.

20. 61.64 is 67% of _____.

Find the percentages.

21. 18.3% of 76 = _13.908_

22. 8.32% of 63 = _____

23. $12\frac{1}{2}$% of 80 = _____

24. 38.2% of 88 = _____

25. $33\frac{1}{3}$% of 72 = _____

26. 7.64% of 32 = _____

Fraction to percent

Complete each of the following.

1. $\frac{3}{5}$ is __60__ percent.

2. $\frac{5}{8}$ is _____ percent.

3. $\frac{3}{4}$ is _____ percent.

4. $\frac{2}{5}$ is _____ percent.

5. $\frac{7}{10}$ is _____ percent.

6. $\frac{1}{2}$ is _____ percent.

7. $\frac{7}{8}$ is _____ percent.

8. $\frac{1}{3}$ is _____ percent.

9. $\frac{2}{3}$ is _____ percent.

10. $\frac{1}{4}$ is _____ percent.

11. 7 is _____ percent of 10.

12. 9 is _____ percent of 27.

13. 15 is _____ percent of 30.

14. 16 is _____ percent of 80.

15. 30 is _____ percent of 50.

16. 17 is _____ percent of 34.

17. 29 is _____ percent of 116.

18. 24 is _____ percent of 120.

19. 45 is _____ percent of 75.

20. 75 is _____ percent of 100.

Complete each of the following.

1. 8 is 25% of _32_.

2. 27 is 75% of _____.

3. 33 is 60% of _____.

4. 56 is $87\frac{1}{2}$% of _____.

5. 42 is 50% of _____.

6. 30 is 75% of _____.

7. 17 is $33\frac{1}{3}$% of _____.

8. 45 is $62\frac{1}{2}$% of _____.

9. 36 is 75% of _____.

10. 44 is 80% of _____.

11. 9.62 is 26% of _____.

12. 8.84 is 52% of _____.

13. 27.36 is 36% of _____.

14. 63.65 is 95% of _____.

15. 37.38 is 89% of _____.

16. 8.48 is 53% of _____.

17. 40.81 is 77% of _____.

18. 22.68 is 27% of _____.

19. 9.44 is 16% of _____.

20. 17.02 is 46% of _____.

Find the percentages.

21. 19.2% of 63 = _12.096_

22. 25% of 84 = _____

23. 53.2% of 29 = _____

24. 0.172% of 69 = _____

25. $66\frac{2}{3}$% of 96 = _____

26. $87\frac{1}{2}$% of 56 = _____

Fraction to percent   Name _____

                             Date _____

Complete each of the following.

1. $\frac{3}{10}$ is __*30*__ percent.   11. 12 is _____ percent of 36.

2. $\frac{5}{8}$ is _____ percent.   12. 24 is _____ percent of 36.

3. $\frac{1}{3}$ is _____ percent.   13. 18 is _____ percent of 27.

4. $\frac{1}{4}$ is _____ percent.   14. 9 is _____ percent of 27.

5. $\frac{2}{3}$ is _____ percent.   15. 25 is _____ percent of 100.

6. $\frac{3}{5}$ is _____ percent.   16. 25 is _____ percent of 50.

7. $\frac{1}{8}$ is _____ percent.   17. 13 is _____ percent of 39.

8. $\frac{3}{4}$ is _____ percent.   18. 26 is _____ percent of 65.

9. $\frac{1}{2}$ is _____ percent.   19. 13 is _____ percent of 104.

10. $\frac{9}{10}$ is _____ percent.   20. 21 is _____ percent of 70.

57

Finding the whole
Finding percentage

Name _____

Date _____

Complete each of the following.

1. 18 is 40% of _45_.

2. 30 is $37\frac{1}{2}$% of _____.

3. 56 is 50% of _____.

4. 51 is 75% of _____.

5. 92 is 80% of _____.

6. 39 is 30% of _____.

7. 153 is 75% of _____.

8. 13 is 10% of _____.

9. 50 is $33\frac{1}{3}$% of _____.

10. 120 is $66\frac{2}{3}$% of _____.

11. 15.64 is 17% of _____.

12. 19.43 is 29% of _____.

13. 32.24 is 8% of _____.

14. 15.12 is 27% of _____.

15. 69.12 is 64% of _____.

16. 43.18 is 127% of _____.

17. 53.01 is 93% of _____.

18. 34.02 is 54% of _____.

19. 37.8 is 28% of _____.

20. 80.04 is 92% of _____.

Find the percentages.

21. 29.3% of 72 = _21.096_

22. 75% of 84 = _____

23. $33\frac{1}{3}$% of 81 = _____

24. 9.21% of 16 = _____

25. 142% of 61 = _____

26. 82.1% of 65 = _____

Fraction to percent   Name _____

                                        Date _____

Complete each of the following.

1. $\frac{5}{8}$ is _62½_ percent.

2. $\frac{1}{3}$ is _____ percent.

3. $\frac{4}{5}$ is _____ percent.

4. $\frac{1}{10}$ is _____ percent.

5. $\frac{1}{4}$ is _____ percent.

6. $\frac{2}{3}$ is _____ percent.

7. $\frac{7}{20}$ is _____ percent.

8. $\frac{9}{10}$ is _____ percent.

9. $\frac{2}{5}$ is _____ percent.

10. $\frac{3}{8}$ is _____ percent.

11. 28 is _____ percent of 40.

12. 30 is _____ percent of 50.

13. 84 is _____ percent of 105.

14. 27 is _____ percent of 81.

15. 95 is _____ percent of 152.

16. 27 is _____ percent of 45.

17. 44 is _____ percent of 55.

18. 63 is _____ percent of 90.

19. 63 is _____ percent of 180.

20. 210 is _____ percent of 315.

Complete each of the following.

1. 20 is 80% of _25_.

2. 70 is 70% of _____.

3. 32 is 25% of _____.

4. 27 is $37\frac{1}{2}$% of _____.

5. 33 is $33\frac{1}{3}$% of _____.

6. 75 is 50% of _____.

7. 54 is $66\frac{2}{3}$% of _____.

8. 85 is $62\frac{1}{2}$% of _____.

9. 27 is 15% of _____.

10. 135 is $62\frac{1}{2}$% of _____.

11. 25.52 is 22% of _____.

12. 18.24 is 38% of _____.

13. 39.75 is 53% of _____.

14. 57.04 is 92% of _____.

15. 110.2 is 116% of _____.

16. 14.94 is 83% of _____.

17. 13.23 is 9% of _____.

18. 27.3 is 42% of _____.

19. 5.44 is 17% of _____.

20. 55.08 is 27% of _____.

Find the percentages.

21. 62.3% of 65 = _40.495_

22. 114% of 25 = _____

23. 9.31% of 55 = _____

24. 1.81% of 16 = _____

25. 63.2% of 84 = _____

26. 25% of 160 = _____

60

Name _____

Date _____

SAVE 40% ON FAMILY-SIZE SHAMPOO!
$1.00 off the regular price
regular $2.50   **NOW $1.50**
Save 10% to 40% on selected items

Norever Hair Dryer	regular	$20.00	NOW	$17.00
Ontime Travel Alarm	regular	$8.48	NOW	$6.36
Potato Instants	regular	$0.70	NOW	$0.63
Player Pocket Radio	regular	$4.95	NOW	$3.96
Sashimi Color TV	regular	$369.90	NOW	$332.91
Gorgon 110 Camera	regular	$26.95	NOW	$16.17

Complete the following table.

	item	regular price	sale price	amount saved	percent saved
1.	Family-Size Shampoo	$2.50	$1.50	$1.00	40%
2.	Norever Hair Dryer				
3.	Ontime Travel Alarm				
4.	Potato Instants				
5.	Player Pocket Radio				
6.	Sashimi Color TV				
7.	Gorgon 110 Camera				

Name _____

Date _____

SAVE 20% ON WARREN LABELS!
$1.19 off the regular price
regular $5.95   NOW $4.76

Columbus	regular	$6.95	NOW	$4.17
MTH	regular	$7.95	NOW	$6.36
Orange	regular	$6.50	NOW	$5.46
Gray Note	regular	$6.40	NOW	$5.44
Dutch Grammar	regular	$11.25	NOW	$9.00
Common Singers	regular	$8.95	NOW	$7.16

Complete the following table.

	item	regular price	sale price	amount saved	percent saved
1.	Warren	$5.95	$4.76	$1.19	20%
2.	Columbus				
3.	MTH				
4.	Orange				
5.	Gray Note				
6.	Dutch Grammar				
7.	Common Singers				

Name _____

Date _____

SAVE 20% ON WOMEN'S JEANS!
$2.40 off the regular price

regular $12.00   **NOW $9.60**

Save 20% to 35% on selected items

Knit Tops	regular	$8.00	NOW	$6.40
Carry-Alls	regular	$13.00	NOW	$9.49
Men's Dress Shirts	regular	$10.00	NOW	$7.90
Men's Slacks	regular	$14.00	NOW	$10.08
Casual Socks	regular	$1.50	NOW	$0.99
Sports Socks	regular	$1.60	NOW	$1.20

Complete the following table.

	item	regular price	sale price	amount saved	percent saved
1.	Women's Jeans	$12.00	$9.60	$2.40	20%
2.	Knit Tops				
3.	Carry-Alls				
4.	Men's Dress Shirts				
5.	Men's Slacks				
6.	Casual Socks				
7.	Sports Socks				

Name _____

Date _____

SAVE 50% ON STEREO HEADPHONES!
$3.49 off the regular price
regular $6.98   **NOW $3.49**
Save 15% to 50% on selected items

AM/FM, 8-Track Record Player	regular	$119.00	NOW	$98.77
Portable Cassette Record	regular	$29.00	NOW	$18.27
AM/FM with Cassette	regular	$49.00	NOW	$37.73
AM Pocket Radio	regular	$9.00	NOW	$7.65
C-60 Cassette 2-Pack	regular	$2.50	NOW	$1.50
AM/FM, Cassette Record Player	regular	$135.00	NOW	$108.00

Complete the following table.

	item	regular price	sale price	amount saved	percent saved
1.	Stereo Headphones	$6.98	$3.49	$3.49	50%
2.	8-Track Record Player				
3.	Cassette Recorder				
4.	AM/FM with Cassette				
5.	AM Pocket Radio				
6.	Cassette 2-Pack				
7.	Cassette Record Player				

Understanding sales

Name _____

Date _____

SAVE 20% ON TIARA COLOR TV!
$105 off the regular price
regular $525   **NOW $420**
Save 15% to 25% on selected items

26'' Tiara Color	regular	$525.00	NOW	$420.00
24'' Tiara Color	regular	$479.00	NOW	$397.57
26'' Sonya Color	regular	$719.00	NOW	$611.15
24'' Sonya Color	regular	$629.00	NOW	$528.36
18'' Sonya Color	regular	$499.00	NOW	$394.21
24'' Sonya B&W	regular	$299.00	NOW	$239.20
20'' Sonya B&W	regular	$279.00	NOW	$217.62

Complete the following table.

	item	regular price	sale price	amount saved	percent saved
1.	26" Tiara Color	$525.00	$420.00	$105.00	20%
2.	24" Tiara Color				
3.	26" Sonya Color				
4.	24" Sonya Color				
5.	18" Sonya Color				
6.	24" Sonya B&W				
7.	20" Sonya B&W				

ANSWERS

Page 1 **1.** 24/100, 6/25, 0.24 **2.** 56/100, 14/25, 0.56 **3.** 92/100, 23/25, 0.92 **4.** 105/100, 21/20, 1.05 **5.** 69/100, 69/100, 0.69 **6.** 53/100, 53/100, 0.53 **7.** 28/100, 7/25, 0.28 **8.** 75/100, 3/4, 0.75 **9.** 25/100, 1/4, 0.25 **10.** 50/100, 1/2, 0.50 **11.** $33\frac{1}{3}$/100, 1/3, $0.33\frac{1}{3}$ **12.** 32/100, 8/25, 0.32 **13.** 145/100, 29/20, 1.45 **14.** $62\frac{1}{2}$/100, 5/8, 0.625 **15.** 176/100, 44/25, 1.76 **16.** 225/100, 9/4, 2.25 **17.** 95/100, 19/20, 0.95 **18.** 42/100, 21/50, 0.42 **19.** 265/100, 53/20, 2.65 **20.** 67/100, 67/100, 0.67

Page 2 **1.** 1/2, 0.50, 50% **2.** 1/4, 0.25, 25% **3.** 7/4, 1.75, 175% **4.** 7/5, 1.40, 140% **5.** 12/5, 2.40, 240% **6.** 17/10, 1.70, 170% **7.** 11/20, 0.55, 55% **8.** 27/20, 1.35, 135% **9.** 37/20, 1.85, 185% **10.** 51/25, 2.04, 204% **11.** 5/4, 1.25, 125% **12.** 3/25, 0.12, 12% **13.** 14/25, 0.56, 56% **14.** 22/25, 0.88, 88% **15.** 24/25, 0.96, 96% **16.** 36/25, 1.44, 144% **17.** 44/25, 1.76, 176% **18.** 53/25, 2.12, 212% **19.** 1/5, 0.20, 20% **20.** 9/20, 0.45, 45%

Page 3 **1.** 68/100, 17/25, 0.68 **2.** 73/100, 73/100, 0.73 **3.** 46/100, 23/50, 0.46 **4.** 29/100, 29/100, 0.29 **5.** 50/100, 1/2, 0.50 **6.** 34/100, 17/50, 0.34 **7.** 26/100, 13/50, 0.26 **8.** 81/100, 81/100, 0.81 **9.** 49/100, 49/100, 0.49 **10.** 56/100, 14/25, 0.56 **11.** 59/100, 59/100, 0.59 **12.** 98/100, 49/50, 0.98 **13.** 156/100, 39/25, 1.56 **14.** 216/100, 54/25, 2.16 **15.** 44/100, 11/25, 0.44 **16.** 114/100, 57/50, 1.14 **17.** 29/100, 29/100, 0.29 **18.** 32/100, 8/25, 0.32 **19.** 67/100, 67/100, 0.67 **20.** $37\frac{1}{2}$/100, 3/8, 0.375

Page 4 **1.** 49/20, 2.45, 245% **2.** 17/20, 0.85, 85% **3.** 9/5, 1.80, 180% **4.** 4/25, 0.16, 16% **5.** 6/5, 1.20, 120% **6.** 27/10, 2.70, 270% **7.** 76/25, 3.04, 304% **8.** 31/20, 1.55, 155% **9.** 7/20, 0.35, 35% **10.** 7/10, 0.70, 70% **11.** 9/4, 2.25, 225% **12.** 3/2, 1.50, 150% **13.** 33/25, 1.32, 132% **14.** 44/25, 1.76, 176% **15.** 8/25, 0.32, 32% **16.** 13/25, 0.52, 52% **17.** 32/25, 1.28, 128% **18.** 1/5, 0.20, 20% **19.** 3/10, 0.30, 30% **20.** 23/10, 2.30, 230%

Page 5 **1.** 35/100, 7/20, 0.35 **2.** 60/100, 3/5, 0.60 **3.** 42/100, 21/50, 0.42 **4.** 28/100, 7/25, 0.28 **5.** 39/100, 39/100, 0.39 **6.** $37\frac{1}{2}$/100, 3/8, 0.375 **7.** 40/100, 2/5, 0.40 **8.** 76/100, 19/25, 0.76 **9.** 53/100, 53/100, 0.53 **10.** 75/100, 3/4, 0.75 **11.** 65/100, 13/20, 0.65 **12.** $33\frac{1}{3}$/100, 1/3, $0.33\frac{1}{3}$ **13.** 72/100, 18/25, 0.72 **14.** 150/100, 3/2, 1.50 **15.** 47/100, 47/100, 0.47 **16.** 12/100, 3/25, 0.12 **17.** 175/100, 7/4, 1.75 **18.** 67/100, 67/100, 0.67 **19.** $12\frac{1}{2}$/100, 1/8, 0.125 **20.** 85/100, 17/20, 0.85

Page 6 **1.** 4/5, 0.80, 80% **2.** 19/20, 0.95, 95% **3.** 51/20, 2.55, 255% **4.** 11/10, 1.10, 110% **5.** 21/20, 1.05, 105% **6.** 43/20, 2.15, 215% **7.** 42/25, 1.68, 168% **8.** 28/25, 1.12, 112% **9.** 11/25, 0.44, 44% **10.** 34/25, 1.36, 136% **11.** 11/4, 2.75, 275% **12.** 1/4, 0.25, 25% **13.** 1/20, 0.05, 5% **14.** 21/25, 0.84, 84% **15.** 47/25, 1.88, 188% **16.** 9/4, 2.25, 225% **17.** 5/2, 2.50, 250% **18.** 8/5, 1.60, 160% **19.** 13/20, 0.65, 65% **20.** 11/5, 2.20, 220%

Page 7 **1.** 102/100, 51/50, 1.02 **2.** 53/100, 53/100, 0.53 **3.** 30/100, 3/10, 0.30 **4.** 36/100, 9/25, 0.36 **5.** 69/100, 69/100, 0.69 **6.** 20/100, 1/5, 0.20 **7.** 5/100, 1/20, 0.05 **8.** 38/100, 19/50, 0.38 **9.** 25/100, 1/4, 0.25 **10.** 84/100, 21/25, 0.84 **11.** 62/100, 31/50, 0.62 **12.** 115/100, 23/20, 1.15 **13.** 48/100, 12/25, 0.48 **14.** 34/100, 17/50, 0.34 **15.** 81/100, 81/100, 0.81 **16.** $66\frac{2}{3}$/100, 2/3, $0.66\frac{2}{3}$ **17.** 325/100, 13/4, 3.25 **18.** $37\frac{1}{2}$/100, 3/8, 0.375 **19.** 155/100, 31/20, 1.55 **20.** $12\frac{1}{2}$/100, 1/8, 0.125

Page 8 **1.** 1/5, 0.20, 20% **2.** 7/5, 1.40, 140% **3.** 3/2, 1.50, 150% **4.** 21/10, 2.10, 210% **5.** 29/20, 1.45, 145% **6.** 23/20, 1.15, 115% **7.** 19/20, 0.76, 76% **8.** 44/25, 1.76, 176% **9.** 2/25, 0.08, 8% **10.** 3/4, 0.75, 75% **11.** 7/4, 1.75, 175% **12.** 9/10, 0.90, 90% **13.** 9/5, 1.80, 180% **14.** 41/20, 2.05, 205% **15.** 37/25, 1.48, 148% **16.** 49/25, 1.96, 196% **17.** 11/50, 0.22, 22% **18.** 6/25, 0.24, 24% **19.** 12/5, 2.40, 240% **20.** 33/20, 1.65, 165%

Page 9 **1.** 28/100, 7/25, 0.28 **2.** 41/100, 41/100, 0.41 **3.** 53/100, 53/100, 0.53 **4.** 100/100, 1/1, 1.00 or 1 **5.** 48/100, 12/25, 0.48 **6.** 32/100, 8/25, 0.32 **7.** 81/100, 81/100, 0.81 **8.** 40/100, 2/5, 0.4 **9.** 7/100, 7/100, 0.07 **10.** 95/100, 19/20, 0.95 **11.** 82/100, 41/50, 0.82 **12.** 75/100, 3/4, 0.75 **13.** 42/100, 21/50, 0.42 **14.** 250/100, 5/2, 2.5 **15.** 125/100, 5/4, 1.25 **16.** $33\frac{1}{3}$/100, 1/3, $0.33\frac{1}{3}$ **17.** 145/100, 29/20, 1.45 **18.** $12\frac{1}{2}$/100, 1/8, 0.125 **19.** 112/100, 28/25, 1.12 **20.** 43/100, 43/100, 0.43

Page 10 **1.** 1/4, 0.25, 25% **2.** 2/5, 0.4, 40% **3.** 1/10, 0.1, 10% **4.** 13/10, 1.3, 130% **5.** 17/20, 0.85, 85% **6.** 5/4, 1.25, 125% **7.** 4/5, 0.8, 80% **8.** 9/5, 1.8, 180% **9.** 37/20, 1.85, 185% **10.** 1/25, 0.04, 4% **11.** 9/25, 0.36, 36% **12.** 26/25, 1.04, 104% **13.** 12/25, 0.48, 48% **14.** 23/10, 2.3, 230% **15.** 31/25, 1.24, 124% **16.** 16/25, 0.64, 64% **17.** 3/20, 0.15, 15% **18.** 11/4, 2.75, 275% **19.** 23/20, 1.15, 115% **20.** 49/20, 2.45, 245%

Page 11 **1.** 21 **2.** 42 **3.** 75 **4.** 14 **5.** 57
6. 9 **7.** 12 **8.** 28 **9.** 54 **10.** 50 **11.** 51 **12.** 90
13. 33 **14.** 63 **15.** 39 **16.** 21 **17.** 42 **18.** 75
19. 14 **20.** 57 **21.** 9 **22.** 12 **23.** 28 **24.** 54
25. 50 **26.** 51 **27.** 90 **28.** 33 **29.** 63 **30.** 39

Page 12 **1.** 70 **2.** 168 **3.** 161 **4.** 252 **5.** 287
6. 70 **7.** 168 **8.** 161 **9.** 252 **10.** 287 **11.** 4.68
12. 14.28 **13.** 55 **14.** 8.96 **15.** 85.2 **16.** 30.24
17. 3.51 **18.** 24.84 **19.** 60.39 **20.** 65.65

Page 13 **1.** 12 **2.** 46 **3.** 72 **4.** 35 **5.** 34
6. 8 **7.** 16 **8.** 48 **9.** 68 **10.** 61 **11.** 12 **12.** 76
13. 20 **14.** 120 **15.** 160 **16.** 12 **17.** 46 **18.** 72
19. 35 **20.** 34 **21.** 8 **22.** 16 **23.** 48 **24.** 68
25. 61 **26.** 12 **27.** 76 **28.** 20 **29.** 120 **30.** 160

Page 14 **1.** 36 **2.** 99 **3.** 261 **4.** 180 **5.** 225
6. 36 **7.** 99 **8.** 261 **9.** 180 **10.** 225 **11.** 33.8
12. 180.34 **13.** 131.56 **14.** 45.39 **15.** 112.05
16. 22.08 **17.** 24.96 **18.** 93.96 **19.** 54.34
20. 110.4

Page 15 **1.** 21 **2.** 60 **3.** 168 **4.** 129 **5.** 273
6. 13 **7.** 48 **8.** 33 **9.** 94 **10.** 128 **11.** 27
12. 36 **13.** 39 **14.** 12 **15.** 69 **16.** 21 **17.** 60
18. 168 **19.** 129 **20.** 273 **21.** 13 **22.** 48
23. 33 **24.** 94 **25.** 128 **26.** 27 **27.** 36 **28.** 39
29. 12 **30.** 69

Page 16 **1.** 136 **2.** 68 **3.** 340 **4.** 323 **5.** 408
6. 136 **7.** 68 **8.** 340 **9.** 323 **10.** 408 **11.** 34.04
12. 15.81 **13.** 19.92 **14.** 112.75 **15.** 65.66
16. 60.84 **17.** 48.36 **18.** 31.46 **19.** 82.08
20. 16.77

Page 17 **1.** 10 **2.** 4. **3.** 18 **4.** 36 **5.** 29
6. 49 **7.** 21 **8.** 89 **9.** 107 **10.** 154 **11.** 70
12. 189 **13.** 245 **14.** 294 **15.** 371 **16.** 10
17. 4 **18.** 18 **19.** 36 **20.** 29 **21.** 49 **22.** 21
23. 89 **24.** 107 **25.** 154 **26.** 70 **27.** 189
28. 245 **29.** 294 **30.** 371

Page 18 **1.** 45 **2.** 72 **3.** 198 **4.** 135 **5.** 270
6. 45 **7.** 72 **8.** 198 **9.** 135 **10.** 270 **11.** 31.68
12. 30.21 **13.** 31.68 **14.** 48.26 **15.** 46.24
16. 55.08 **17.** 23.56 **18.** 24.94 **19.** 20.52
20. 44.08

Page 19 **1.** 28 **2.** 60 **3.** 48 **4.** 84 **5.** 180
6. 40 **7.** 50 **8.** 75 **9.** 45 **10.** 155 **11.** 90
12. 45 **13.** 126 **14.** 207 **15.** 315 **16.** 28
17. 60 **18.** 48 **19.** 84 **20.** 180 **21.** 40
22. 50 **23.** 75 **24.** 45 **25.** 155 **26.** 90
27. 45 **28.** 126 **29.** 207 **30.** 315

Page 20 **1.** 28 **2.** 126 **3.** 259 **4.** 203 **5.** 406
6. 28 **7.** 126 **8.** 259 **9.** 203 **10.** 406 **11.** 19.52

12. 41.34 **13.** 46.62 **14.** 15.54 **15.** 54.02
16. 29.7 **17.** 20.44 **18.** 40.48 **19.** 44.16
20. 51.52

Page 21 **1.** 50 **2.** $33\frac{1}{3}$ **3.** $66\frac{2}{3}$ **4.** 25 **5.** 75
6. 20 **7.** 40 **8.** 60 **9.** 80 **10.** 160 **11.** $12\frac{1}{2}$
12. $37\frac{1}{2}$ **13.** $62\frac{1}{2}$ **14.** $87\frac{1}{2}$ **15.** $112\frac{1}{2}$ **16.** 70
17. 30 **18.** 90 **19.** 125 **20.** $137\frac{1}{2}$ **21.** 10
22. 30 **23.** 49 **24.** 36 **25.** 60 **26.** 5 **27.** 9
28. 8 **29.** 90 **30.** 22 **31.** 24 **32.** 20 **33.** 108
34. 154 **35.** 49 **36.** 535

Page 22 **1.** 0.48 **2.** 0.92 **3.** 0.56 **4.** 0.12
5. 0.59 **6.** 0.17 **7.** 0.87 **8.** 0.42 **9.** 0.31
10. 1.25 **11.** 1.67 **12.** 1.05 **13.** 1.10 **14.** 0.007
15. 0.03 **16.** 2.50 **17.** 7.00 **18.** 0.01 **19.** 0.175
20. 0.093 **21.** 13.92 **22.** 69. 92 **23.** 11.16
24. 51.33 **25.** 48.3 **26.** 5.95 **27.** 37.41
28. 23.25 **29.** 115 **30.** 32.5 **31.** 70.14
32. 28.35 **33.** 2.268 **34.** 5.12 **35.** 3.85
36. 8.82

Page 23 **1.** 62.5 **2.** $33\frac{1}{3}$ **3.** 25 **4.** 137.5 **5.** 80
6. $66\frac{2}{3}$ **7.** 125 **8.** 37.5 **9.** 375 **10.** 340 **11.** 75
12. 87.5 **13.** 162.5 **14.** 70 **15.** $22\frac{2}{9}$ **16.** 180
17. 90 **18.** 60 **19.** $166\frac{2}{3}$ **20.** 30 **21.** 38 **22.** 85
23. 81 **24.** 255 **25.** 147 **26.** 49 **27.** 212
28. 18 **29.** 50 **30.** 56 **31.** 224 **32.** 99 **33.** 105
34. 177 **35.** 72 **36.** 161

Page 24 **1.** 0.36 **2.** 0.57 **3.** 0.46 **4.** 0.08
5. 0.13 **6.** 0.31 **7.** 0.19 **8.** 0.16 **9.** 0.13
10. 1.32 **11.** 0.27 **12.** 0.054 **13.** 0.037
14. 0.081 **15.** 0.163 **16.** 0.082 **17.** 0.174
18. 0.136 **19.** 0.159 **20.** 0.281 **21.** 15.48
22. 15.96 **23.** 14.72 **24.** 18.59 **25.** 13.76
26. 8.99 **27.** 14.25 **28.** 64.68 **29.** 2.106
30. 1.776 **31.** 2.214 **32.** 3.402 **33.** 9.112
34. 8.586 **35.** 25.852 **36.** 1.98

Page 25 **1.** 37.5 **2.** 125 **3.** 160 **4.** $33\frac{1}{3}$
5. 325 **6.** 60 **7.** 87.5 **8.** 25 **9.** $66\frac{2}{3}$ **10.** 70
11. 137.5 **12.** 50 **13.** 40 **14.** 62.5 **15.** 80
16. 20 **17.** 30 **18.** $166\frac{2}{3}$ **19.** 75 **20.** $55\frac{5}{9}$
21. 91 **22.** 55 **23.** 81 **24.** 38 **25.** 60 **26.** 112
27. 8 **28.** 33 **29.** 98 **30.** 108 **31.** 31 **32.** 80
33. 168 **34.** 75 **35.** 135 **36.** 41

Page 26 **1.** 0.26 **2.** 0.05 **3.** 0.16 **4.** 0.82
5. 0.08 **6.** 0.12 **7.** 0.29 **8.** 0.32 **9.** 0.09
10. 0.17 **11.** 0.069 **12.** 0.0762 **13.** 0.0031
14. 8.10 **15.** 0.096 **16.** 0.032 **17.** 0.0514
18. 0.0023 **19.** 0.003 **20.** 1.14 **21.** 22.62
22. 15.36 **23.** 13.12 **24.** 40.24 **25.** 348.3
26. 8.04 **27.** 16.24 **28.** 26.24 **29.** 21.87
30. 18.02 **31.** 2.208 **32.** 3.7522 **33.** 0.1219
34. 8.193 **35.** 33.06 **36.** 0.2412

Page 27   1. 90   2. 12.5   3. 130   4. 125   5. $11\frac{1}{9}$ 6. 225   7. 25   8. 75   9. 37.5   10. 87.5   11. 40   12. 212.5   13. 62.5   14. 325   15. 70   16. 140   17. 187.5   18. 50   19. 180   20. $122\frac{2}{9}$ 21. 6   22. 36   23. 21   24. 28   25. 75   26. 22   27. 126   28. 117   29. 11   30. 47   31. 153   32. 207   33. 195   34. 7   35. 155   36. 535

Page 28   1. 0.22   2. 0.13   3. 0.72   4. 0.81   5. 0.24   6. 0.69   7. 0.52   8. 0.55   9. 0.32   10. 1.37   11. 0.172   12. 0.181   13. 0.053   14. 1.13   15. 0.151   16. 0.092   17. 0.0035   18. 0.014   19. 0.095   20. 0.081   21. 17.82   22. 8.45   23. 9.88   24. 51.03   25. 14.4   26. 12.24   27. 20.16   28. 5.504   29. 2.715   30. 0.294   31. 11.476   32. 5.244   33. 2.1   34. 0.392   35. 63.28   36. 11.591

Page 29   1. 40   2. $66\frac{2}{3}$ 3. 125   4. 140   5. 70   6. $166\frac{2}{3}$ 7. 37.5   8. 150   9. 60   10. 75   11. 80   12. $33\frac{1}{3}$ 13. 50   14. 12.5   15. 20   16. 30   17. 175   18. $233\frac{1}{3}$ 19. $11\frac{1}{9}$ 20. 130   21. 45   22. 63   23. 135   24. 33   25. 65   26. 52   27. 50   28. 135   29. 119   30. 12   31. 81   32. 161   33. 36   34. 10   35. 22   36. 45

Page 30   1. 0.42   2. 0.35   3. 0.58   4. 0.07   5. 0.16   6. 0.83   7. 0.09   8. 0.72   9. 0.77   10. 0.89   11. 0.122   12. 0.087   13. 1.15   14. 0.0032   15. 0.069   16. 0.035   17. 0.129   18. 0.052   19. 0.141   20. 0.0003   21. 39.48   22. 9.45   23. 36.54   24. 167.09   25. 64.8   26. 33.84   27. 65.17   28. 9.272   29. 29.145   30. 31.05   31. 0.2976   32. 7.038   33. 0.366   34. 6.837   35. 8.037   36. 0.345

Page 31   1. 3.84   2. 19   3. 12.09   4. 17   5. 30.71   6. 11   7. 7.381   8. 57   9. 8.487   10. 124   11. 96   12. 1.1096   13. 45   14. 10.152   15. 43   16. 1.26   17. 342   18. 62.568   19. 110   20. 5.916

Page 32   1. 50%   2. 60%   3. 25%   4. 10%   5. 80%   6. 75%   7. 30%   8. 62.5%   9. 40%   10. 87.5%   11. 80%   12. 70%   13. 50%   14. 75%   15. 62.5%   16. 40%   17. 30%   18. 25%   19. 60%   20. 50%

Page 33   1. 4.76   2. 50.16   3. 23   4. 13.871   5. 1.792   6. 28   7. 7.524   8. 38.64   9. 72   10. 81.9   11. 288   12. 12.159   13. 31.05   14. 281   15. 12.16   16. 22.62   17. 260   18. 55.61   19. 49.383   20. 11.658

Page 34   1. $33\frac{1}{3}$%   2. 75%   3. 37.5%   4. 60%   5. 50%   6. 80%   7. $66\frac{2}{3}$%   8. 62.5%   9. 25%   10. 70%   11. 60%   12. 62.5%   13. 90%   14. 50%   15. 25%   16. 70%   17. 75%   18. 37.5%   19. 30%   20. 40%

Page 35   1. 111.93   2. 16438   3. 187.11   4. 63   5. 93.48   6. 882   7. 39.06   8. 348   9. 2.432   10. 1689   11. 5.561   12. 99   13. 14.56   14. 8   15. 91.28   16. 333   17. 39.48   18. 854   19. 333   20. 17.48

Page 36   1. 90%   2. 62.5%   3. $66\frac{2}{3}$%   4. 60%   5. 50%   6. $33\frac{1}{3}$%   7. 37.5%   8. 75%   9. 15%   10. $58\frac{1}{3}$%   11. 80%   12. 40%   13. 90%   14. 110%   15. 50%   16. 150%   17. 60%   18. 200%   19. 30%   20. 20%

Page 37   1. 10.956   2. 83   3. 246   4. 775   5. 96.74   6. 157.08   7. 40.32   8. 2253   9. 408.28   10. 201   11. 46   12. 996   13. 4.992   14. 225   15. 12   16. 2.563   17. 500   18. 38.28   19. 334   20. 18

Page 38   1. 60%   2. 30%   3. 75%   4. $33\frac{1}{3}$%   5. 87.5%   6. 62.5%   7. $28\frac{4}{7}$%   8. 70%   9. $66\frac{2}{3}$%   10. 90%   11. 50%   12. 300%   13. 150%   14. 120%   15. 75%   16. 80%   17. $33\frac{1}{3}$%   18. $33\frac{1}{3}$%   19. 70%   20. 62.5%

Page 39   1. 35.91   2. 993   3. 6.642   4. 4632   5. 88.56   6. 421   7. 2.881   8. 1.728   9. 543   10. 202.34   11. 532   12. 24.7   13. 33   14. 16   15. 205.74   16. 222   17. 47.74   18. 500   19. 20.193   20. 581

Page 40   1. 60%   2. $33\frac{1}{3}$%   3. 50%   4. 62.5%   5. 80%   6. 25%   7. $66\frac{2}{3}$%   8. 37.5%   9. 70%   10. 35%   11. 50%   12. $33\frac{1}{3}$%   13. 25%   14. 25%   15. 50%   16. 75%   17. 30%   18. 150%   19. $66\frac{2}{3}$%   20. 50%

Page 41   1. savings $3.89, sale price $35.01   2. savings $14.18, sale price $21.27   3. savings $5.49, sale price $16.47   4. savings $11.40, sale price $17.10   5. savings $5.93, sale price $13.82   6. savings $11.00, sale price $16.50

Page 42   1. savings $3.65, sale price $14.60   2. savings $5.58, sale price $11.17   3. savings $9.80, sale price $14.70   4. savings $1.83, sale price $16.47   5. savings $3.39, sale price $13.56   6. savings $2.05, sale price $14.35   7. savings $2.15, sale price $12.15   8. savings $0.79, sale price $7.11

Page 43   1. savings $11.35, sale price $34.05   2. savings $6.45, sale price $15.05   3. savings $5.70, sale price $22.80   4. savings $4.30, sale price $17.20   5. savings $3.90, sale price $22.10   6. savings $8.10, sale price $18.90

Page 44 1. savings $5.00, sale price $10.00
2. savings $4.87, sale price $14.61 3. savings
$2.00, sale price $14.00 4. savings $5.10, sale
price $7.65 5. savings $3.30, sale price $13.20
6. savings $1.61, sale price $14.49 7. savings
$4.00, sale price $8.00 8. savings $1.00, sale
price $7.00

Page 45 1. savings $23.98, sale price $35.97
2. savings $7.20, sale price $21.60 3. savings
$7.77, sale price $18.13 4. savings $2.82, sale
price $15.98 5. savings $1.98, sale price $17.82
6. savings $8.00, sale price $24.00

Page 46 1. savings $1.40, sale price $12.60
2. savings $7.95, sale price $18.55 3. savings
$8.00, sale price $16.00 4. savings $6.60, sale
price $9.90 5. savings $4.35, sale price $13.05
6. savings $6.00, sale price $12.00 7. savings
$2.59, sale price $10.36 8. savings $3.00, sale
price $6.00

Page 47 1. savings $7.99, sale price $31.96
2. savings $1.75, sale price $15.75 3. savings
$8.10, sale price $24.30 4. savings $5.86, sale
price $17.58 5. savings $5.10, sale price $20.40
6. savings $5.26, sale price $21.04

Page 48 1. savings $9.58, sale price $14.37
2. savings $3.24, sale price $12.96 3. savings
$5.85, sale price $17.55 4. savings $0.92, sale
price $8.28 5. savings $2.16, sale price $12.24
6. savings $5.55, sale price $12.95 7. savings
$2.50, sale price $10.00 8. savings $1.79, sale
price $7.16

Page 49 1. savings $9.99, sale price $39.96
2. savings $8.00, sale price $24.00 3. savings
$2.85, sale price $25.65 4. savings $2.15, sale
price $19.35 5. savings $6.86, sale price $20.58
6. savings $5.25, sale price $29.75

Page 50 1. savings $3.50, sale price $14.00
2. savings $4.60, sale price $13.80 3. savings
$4.40, sale price $17.60 4. savings $4.38, sale
price $6.57 5. savings $2.79, sale price $15.81
6. savings $3.24, sale price $12.96 7. savings
$3.60, sale price $10.80 8. savings $1.00, sale
price $9.00

Page 51 1. 50 2. 40 3. 12.5 4. $33\frac{1}{3}$ 5. 80
6. 62.5 7. $66\frac{2}{3}$ 8. 75 9. 30 10. 15 11. 50
12. 60 13. 20 14. 25 15. 50 16. $33\frac{1}{3}$ 17. 75
18. 20 19. 25 20. $33\frac{1}{3}$

Page 52 1. 48 2. 36 3. 50 4. 90 5. 112
6. 45 7. 95 8. 1100 9. 120 10. 105 11. 32
12. 48 13. 105 14. 16 15. 90 16. 27 17. 36

18. 81 19. 52 20. 14 21. 4.794 22. 10.205
23. 32 24. 12.028 25. 12.702 26. 9

Page 53 1. 50 2. 80 3. 75 4. 30 5. 40
6. 25 7. 90 8. 12.5 9. 20 10. 62.5 11. 60%
12. 70% 13. 50% 14. 80% 15. 87.5%
16. 75% 17. 50% 18. 75% 19. 25% 20. 75%

Page 54 1. 8 2. 32 3. 10 4. 28 5. 18
6. 45 7. 52 8. 64 9. 45 10. 144 11. 19
12. 32 13. 14 14. 86 15. 24 16. 81 17. 46
18. 57 19. 24 20. 92 21. 13.908 22. 5.2416
23. 10 24. 33.616 25. 24 26. 2.4448

Page 55 1. 60 2. 62.5 3. 75 4. 40 5. 70
6. 50 7. 87.5 8. $33\frac{1}{3}$ 9. $66\frac{2}{3}$ 10. 25 11. 70
12. $33\frac{1}{3}$ 13. 50 14. 20 15. 60 16. 50
17. 25 18. 20 19. 60 20. 75

Page 56 1. 32 2. 36 3. 55 4. 64 5. 84
6. 40 7. 51 8. 72 9. 48 10. 55 11. 37.
12. 17 13. 76 14. 67 15. 42 16. 16 17. 53
18. 84 19. 59 20 37 21. 12.096 22. 21
23. 15.428 24. 0.11868 25. 64 26. 49

Page 57 1. 30. 2. 62.5 3. $33\frac{1}{3}$ 4. 25 5. $66\frac{2}{3}$
6. 60. 7. 12.5 8. 75 9. 50 10. 90 11. $33\frac{1}{3}$
12. $66\frac{2}{3}$ 13. $66\frac{2}{3}$ 14. $33\frac{1}{3}$ 15. 25 16. 50
17. $33\frac{1}{3}$ 18. 40 19. 12.5 20. 30

Page 58 1. 45 2. 80 3. 112 4. 68 5. 115
6. 130 7. 204 8. 130 9. 150 10. 180 11. 92
12. 67 13. 403 14. 56 15. 108 16. 34 17. 57
18. 63 19. 135 20. 87 21. 21.096 22. 63
23. 27 24. 1.4736 25. 86.62 26. 53.365

Page 59 1. 62.5 2. $33\frac{1}{3}$ 3. 80 4. 10 5. 25
6. $66\frac{2}{3}$ 7. 35 8. 90 9. 40 10. 37.5 11. 70
12. 60 13. 80 14. $33\frac{1}{3}$ 15. 62.5 16. 60 17. 80
18. 70 19. 35 20. $66\frac{2}{3}$

Page 60 1. 25 2. 100 3. 128 4. 72 5. 99
6. 150 7. 81 8. 136 9. 180 10. 216 11. 116
12. 48 13. 75 14. 62 15. 95 16. 18 17. 147
18. 65 19. 32 20. 204 21. 40.495 22. 28.5
23. 5.1205 24. 0.2896 25. 53.088 26. 40

Page 61 1. regular price $2.50, sale price $1.50,
amount saved $1.00, percent saved 40%
2. regular price $20.00, sale price $17.00, amount
saved $3.00, percent saved 15% 3. regular price
$8.48, sale price $6.36, amount saved $2.12,
percent saved 25% 4. regular price $0.70, sale
price $0.63, amount saved $0.07, percent saved
10% 5. regular price $4.95, sale price $3.96,
amount saved $0.99, percent saved 20%

69

6. regular price $369.90, sale price $332.91, amount saved $36.99, percent saved 10%   7. regular price $26.95, sale price $16.17, amount saved $10.78, percent saved 40%

Page 62   1. regular price $5.95, sale price $4.76, amount saved $1.19, percent saved 20%
2. regular price $6.95, sale price $4.17, amount saved $2.78, percent saved 40%   3. regular price $7.95, sale price $6.36, amount saved $1.59, percent saved 20%   4. regular price $6.50, sale price $5.46, amount saved $1.04, percent saved 16%   5. regular price $6.40, sale price $5.44, amount saved $0.96, percent saved 15%
6. regular price $11.25, sale price $9.00, amount saved $2.25, percent saved 20%   7. regular price $8.95, sale price $7.16, amount saved $1.79, percent saved 20%

Page 63   1. regular price $12.00, sale price $9.59, amount saved $2.40, percent saved 20%
2. regular price $8.00, sale price $6.40, amount saved $1.60, percent saved 20%   3. regular price $13.00, sale price $9.49, amount saved $3.51, percent saved 27%   4. regular price $10.00, sale price $7.90, amount saved, $2.10, percent saved 21%   5. regular price $14.00, sale price $10.08, amount saved $3.92, percent saved 28%
6. regular price $1.50, sale price $0.99, amount saved $0.51, percent saved 34%   7. regular price

$1.60, sale price $1.20, amount saved $0.40, percent saved 25%

Page 64   1. regular price $6.98, sale price $3.49, amount saved $3.49, percent saved 50%
2. regular price $119.00, sale price $98.77, amount saved $20.23, percent saved 17%   3. regular price $29.00, sale price $18.27, amount saved $10.73, percent saved 37%   4. regular price $49.00, sale price $37.73, amount saved $11.27, percent saved 23%   5. regular price $9.00, sale price $7.65, amount saved $1.35, percent saved 15%
6. regular price $2.50, sale price $1.50, amount saved $1.00, percent saved 40%   7. regular price $135.00, sale price $108.00, amount saved $27.00, percent saved 20%

Page 65   1. regular price $525.00, sale price $420.00, amount saved $105.00, percent saved 20%   2. regular price $479.00, sale price $397.57, amount saved $81.43, percent saved 17%
3. regular price $719.00, sale price $611.15, amount saved $107.85, percent saved 15%   4. regular price $629.00, sale price $528.36, amount saved $100.64, percent saved 16%   5. regular price $499.00, sale price $394.21, amount saved $104.79, percent saved 21%   6. regular price $299.00, sale price $239.20, amount saved $59.80, percent saved 20%   7. regular price $279.00, sale price $217.62. amount saved $61.38, percent saved 22%